U.S. Solar Power Supply *

NATIONAL SIZED SYSTEM*

WITH LONG TERM ENERGY STORAGE*

Provides Power 24 hours per day 365 Days Per Year

Equal To Total U.S. Electrical Demand [DOE 2015]

From Less Than 1.2 Tenths of 1% of U.S. Land Area **

Without Any Conventional Polluting Fossil Fuel or Nuclear Backup

Total Required Solar Area Apportioned Among The Several States Of Arizona, California, Colorado, Nevada, New Mexico, Texas, and Utah: Small Squares Depicted Above: Each Area Is Slightly Less Than 25 Miles by 25 Miles

*Inventor & Author
Randy Ross
RandyRossSolar@Comcast.net
703-606-2909

Ω Six Months Storage at Greater Than 99% Efficiency   ** See Derivation Pages
DOE 2015 Total U.S. Annual Electrical Demand, Public And Private Production 3,911 Billion Kwhrs$_{Electric}$
Efficient Solar Electrical Power At 40% Efficiency** *After* Long Term Energy Storage* REPLACES Conventional Power Plant - Not Just Reduces Its Fuel Use

Table Of Contents

U.S. SOLAR POWER SUPPLY: Description Of Workings

Why Use Common Table Salt For
LONG TERM THERMAL ENERGY STORAGE

The Derivation Of Molten Salt NaCl Storage Capacity

Why Use Sodium Vapor To Transfer Collected Energy:
Sodium Thermal Properties

Overview of the U.S. SOLAR POWER SUPPLY

U.S. SOLAR POWER SUPPLY PILOT PROJECT
Brief Description Of Proposed Project

Frequent Questions About The U.S. SOLAR POWER SUPPLY
Size of Systems, Efficiency, Self Sufficiency, Land Use Requirements
Applications and Ramifications

U.S. SOLAR POWER SUPPLY PILOT PROJECT
Objectives, Technical Areas of Interest And Milestones

U.S. SOLAR POWER SUPPLY PILOT PROJECT
Solar / Electric Conversion Efficiency Summary :
620 Kw$_E$ Peak, 611 Kw$_E$ Net,
Storage: 31 Days No Sun Continuous 24/7 @ ~180 Kw$_E$ = 1
= 33,176 Kwhr$_E$

Detailed Derivation of U.S. Solar Power Supply Pilot Project
Solar to Electric Conversion Efficiency (19 Pages)

U.S. Solar Power Supply City Sized System ** With Six Months Thermal Storage Duration - Summer To Winter At Continuous Average Demand Output

100 Meter Tracking Smooth Surface Dish 1972 Effelsberg: Photograph With Article

U.S. Solar Power Supply City Sized System
With Six Months Long Term Thermal Energy Storage Duration –
At Continuous Average Demand

Output Including Area Needed For 2050 AD Demand Including All Electric Transportation

The Case For Long Term Thermal Energy Storage
i.e. 6 Months Seasonal Storage

The Case For Maximized Solar/Electrical Conversion Efficiency
(It is Almost Rocket Science)

U.S. Solar Power Supply vs. Photovoltaic & Distributed Generation

U.S. Solar Power Supply vs Central Receiver & Parabolic Trough Collectors

Expressions Of Support From: DOE, JPL, INL, University Of Arizona, MIT, DARPA

U.S. Solar Power Supply Land Trust:
Setting Aside Current Government Land For An All Solar America

**All Rights Reserved By Randal Ross 2019

U.S. Solar Power Supply

Efficient Solar Electrical Power Production With *Long Term Storage*

REPLACES Conventional Power Plant – And All Its Infrastructure

And All Its CO_2, Air, Thermal, Land, Water Pollution – And All Its

Coal particulates

Original Invention Disclosed to Denver, Colorado Patent Attorney Ancel Lewis May 14, 1979

The Department Of Energy Has Suppressed This Technology For 40 Years

Description Of Workings Figures 1 & 2. From US Patent 5.685,151

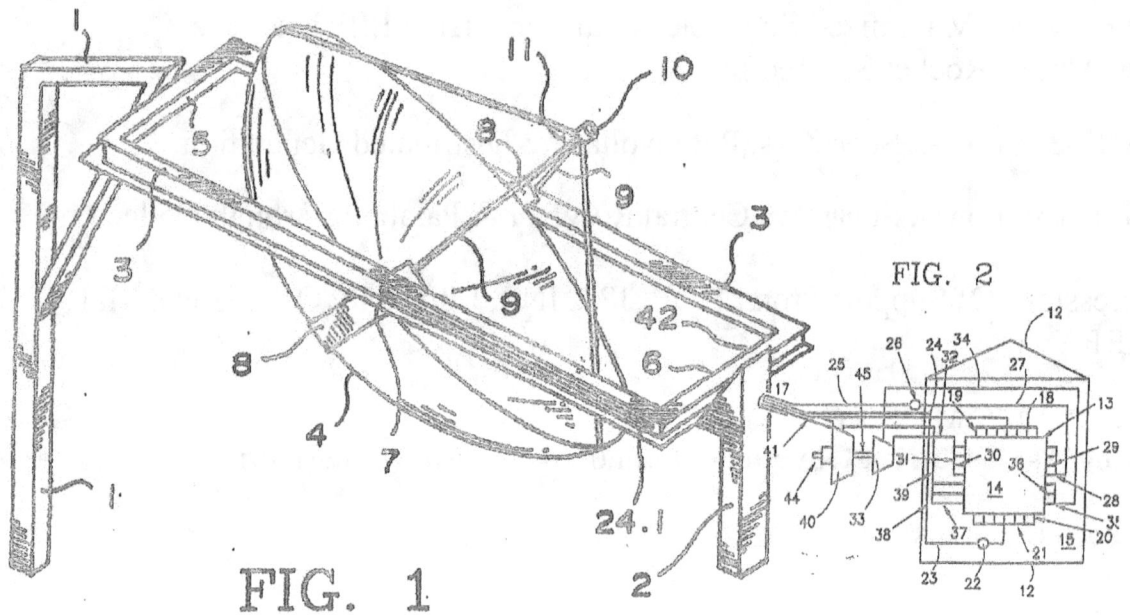

The U.S. Solar Power Supply * utilizes a reflective parabolic reflective dish 4 that tracks the sun suspended by equatorial yoke 3 and focuses its direct rays into a small black body cavity boiler 10 that captures the energy through

several internal reflections. Due to the boiler's small size, radiation convection and conduction losses are small. Regeneratively preheated liquid sodium pumped through tubes that line the cavity to absorb the captured energy isothermally at uniformly high temperature.

As it absorbs energy the sodium vaporizes at 1620°F. and is piped down through 9 boiler support, 7 seasonal adjust support gimbal, 6 inside south support gimbal, to 17 outside south support gimbal to a distribution header 18 where the vapor is channeled through many small tubes 19 in contact with NaCl, common table salt 14 inside a large insulated vessel 13: Where diameter of the reflector 4 is forty-five meters and edge of LONG TERM THERMAL STORAGE 13, a cube, is 10 meters.

Being in contact with the cooler salt, the sodium condenses transferring its energy to the salt which melts at 1486°F storing over 8.286 kilowatt-hours of thermal energy per cubic foot. A second set of tubes 29, and 36 in contact with the molten salt transporting very high pressure water generates 1350°F. supercritical pressure steam 32 suitable for the world's most efficient turbine-generators 33, 40 allowing over 3.103 kilowatt-hours of net electrical generation per cubic foot of molten salt after long term thermal conduction storage losses through tank insulation 15.

City-Sized System Overall Solar/Electrical Efficiency 40.27% including Long Term Storage. See page 38 of 54 page "U.S. SOLAR POWER SUPPLY Solar Electric Conversion Efficiency", a detailed 19 page derivation of the efficiency including very high temperature supercritical pressure turbine with extensive regenerative feed water heating and one reheat cycle in the turbine with exhaust steam expanded down 0.25psi to an average 60°F.

Why Use Common Table Salt For
LONG TERM THERMAL ENERGY STORAGE

The Derivation Of Molten Salt – NaCl - Thermal Storage Capacity

The amount of energy stored in the one cubic foot of molten salt

(NaCl) at 808°C*** = 1486.4°F. in its Heat of Fusion :

Specific Gravity of NaCl = 2.164*** = 2.164 x 1 gr/cc (H2O) = 2.164 gr/cc

cc/cubic foot = 28,317cc/ft³***

Weight Of 1 Cubic Foot Of Salt = 2.164 gr/cc x 28,317cc/f ft³
= 61,277gr/ ft³

61,277gr/ft³ / 453.59gr/pound*** = 135 lbs/ ft³

Specific Heat of Fusion for NaCl = 116.3gr.cal/gr***

116.3gr.cal/gr x 61,277gr/ ft³ of NaCl = 7,126,515 cal/ ft³

0.0000011628 KwHr/Cal*** x 7,126,515 cal/ ft³
= 8.286 KwHr/Ft³ Molten NaCl

Industrial Grade Salt $ 0.04/ pound****

135 lbs/ft3 x $ 0.04/ pound = $ 5.40/ ft³

$5.40/ ft3 Of Salt/ 8.286 KwHr$_T$ / ft³ Molten NaCl = $ 0.65 / Kw Hr$_T$

$ 0.65 / KwHr$_T$ Salt Cost / 37.455978% Solar To Electrical Efficiency*****

= $1.73/KwHr$_E$ Recovered After Long Term Storage (Stored - Recovered Multiple Times As Required - $1.73/KwHr$_E$ Replaces Cost Of Kw$_E$ Capacity At Conventional Plant)

No other material stores this amount of energy so compactly so thermodynamically available at such a low cost.

***Lange's Handbook Of Chemistry 1952 ****Price Quote From International Salt Broker Late 2014
*****
See "U.S. SOLAR POWER SUPPLY PILOT PROJECT
Solar-Electric Conversion Efficiency Summary" contained herein.
See also "U.S. SOLAR POWER SUPPLY* Solar-Electric Conversion Efficiency" a 19 page detailed derivation contained herein

U.S. SOLAR POWER SUPPLY
Efficient Solar Electrical Power Production With *Long Term Storage* REPLACES Conventional Power Plant And Its Strip Mine

Why Use Sodium Vapor
To Transfer Collected Energy : Sodium : Thermal Properties

Sodium Vapor From Solar Boiler Condensing To Liquid Sodium Transferring Energy To Melt Salt (NaCl) In LONG TERM THERMAL STORAGE

Boiling Point Approximate	1620 °F****
Density	46 Lbs/ft3****
Viscosity	4/lbs/Ft/Hr****
Thermal Conductivity	31 Btu/Hr/Ft/°F****
Specific Heat	0.30 Btu/Lb/°F.****
Heat Of Vaporization	1718 Btu/Lb****

$$\frac{3413 \text{ Btu/KwHr}_T \text{ ***}}{1718 \text{ Btu/Lb Condensed Sodium}} = 1.987 \text{ Lbs Mass Flow Na/Kw Hr}_T \text{ Stored In Molten Salt}$$

Mass Flow Per Net Electrical Draw down

It takes only 1.987 Lbs Mass Flow Na/Kw Hr_T Stored =
~ 48 % Thermal-Electric Efficiency *****

= $\frac{4.14 \text{ Lbs Mass Flow Na}}{\text{KwHr}_E}$

Thus only 4.14 Lbs of sodium are vaporized at the solar boiler for every $KwHr_E$ generated including storage losses.

***Lange's Handbook Of Chemistry 1952
****Fast Reactor Technology, M.I.T. Press, 1966, pages 19, 32,33
*****Before Sodium Pump, Clock Drive, External Lighting.
See "U.S. SOLAR POWER SUPPLY PILOT PROJECT
Solar-Electric Conversion Efficiency Summary" contained herein.
See also "U.S. SOLAR POWER SUPPLY
Solar-Electric Conversion Efficiency" a 19 page detailed derivation contained herein.

Overview

During the last 35 years, the Department of Energy has spent 4.6 billion dollars on research, development and demonstration of solar-electrical technology without implementing any system that has LONG TERM ENERGY STORAGE. The capital cost of a solar power system *without* LONG TERM ENERGY STORAGE can only be written off against the fuel savings of the conventional power plant it supplements.

Solar energy derived from these systems merely reduce the conventional power plant's fuel use while the sun shines, or for few additional hours with short term storage, but fails to provide power at night, during extended cloudy periods, and will fall short during those seasons of the year when demand exceeds supply.

DOE 2014

Electrical Production by Renewables in 2012

Power Source	Units in Operation	Power Capacity (GW)	Summer Capacity (GW)	% of Summer Capacity	Capacity Factor	Annual Energy (billion kWh)	% of annual production
Wind	947	59.6	59	5.5	0.272	140.82	3.44
Wood	351	8.5	7.5	0.7	0.575	37.8	0.92
Biomass	1766	5.5	4.8	0.45	0.471	19.82	0.48
Geothermal	197	3.7	2.6	0.24	0.6983	15.56	0.38
Solar	553	3.2	3.2	0.3	0.154	4.33	0.11
Total	7,837	158.7	155.8	14.66	0.362	494.57	12.08

After all these years and billions of dollars of investment just 1/11th of one percent of our electricity is provided by solar generation. Solar is only providing electricity 15.4% of the time. Coal, nuclear and gas (natural, propane) is backing solar up 84.6% of time.

To provide solar power to a given service area using non-storage solar, the utilities and their customers must bear the capital cost of the solar power plant *and* the required conventional backup power plant as well.

However the U.S. SOLAR POWER SUPPLY CITY SIZED SYSTEM with its LONG TERM ENERGY STORAGE *replaces* a conventional power plant, rather than merely reducing the conventional power plant's fuel cost. As a consequence the capital cost of the U.S. SOLAR POWER SUPPLY is written off against the capital cost of the conventional power plant it replaces.

The U.S. SOLAR POWER SUPPLY is a solar-to-electrical technology that utilizes efficient solar energy collection coupled with extended duration (six months) high temperature thermal storage making energy available for the highest temperature and pressure turbines to maximize overall conversion efficiency.

Having LONG TERM ENERGY STORAGE allows the U.S. SOLAR POWER SUPPLY to stand alone without conventional power plant backup *thereby eliminating the capital cost of the conventional power plant*. The U.S. SOLAR POWER SUPPLY CITY SIZED SYSTEM Continuous Rated Output is the power available 24 hours per day 365 days per year.

The U.S. SOLAR POWER SUPPLY PILOT PROJECT allows demonstration of all facets of the proposed system including LONG TERM ENERGY STORAGE on the scale of a one month storage rather than six months for CITY SIZED SYSTEM.

U.S. SOLAR POWER SUPPLY PILOT PROJECT

Proposed:

U.S. SOLAR POWER SUPPLY PILOT PROJECT Demonstrates 31 Days Storage = 133,176 KwHr$_E$ ~ 180 Kw$_E$ No Sun Continuous 24hr/day x 31days. Demonstrates The Smallest Solar Reflector/Land Area Per Electrical Output KwHr$_E$ *After* Long Term Storage Losses: Minimum Working Fluid Mass Flow, Minimum Working Fluid Pumping Losses, Minimum Rankine Cycle Losses Per Net Electrical Output Kwhr$_E$ After *Long Term Storage* Losses: U.S. SOLAR POWER SUPPLY PILOT PROJECT Has Self-Contained Air Cooled Condenser to Eliminate Need For Cooling Water and Thermal Water Pollution

Objectives

Test components, build and operate a pilot U.S. SOLAR POWER SUPPLY PILOT PROJECT to produce 620 Kw$_E$ peak solar-storage-electrical through put:

Prove the ability of the v to gather and store captured solar energy at high efficiency for extended periods of time, storage, as thermal energy stored at high temperature in the LONG TERM ENERGY STORAGE available for the highest efficiency turbine-generators in use today.

Prove the efficacy of sodium as the ultra-high temperature heat transfer medium required for storing thermal energy in molten NaCl making energy available as supercritical pressure steam at or above 1350°F.for the very highest efficiency turbines for overall high solar-to-electrical conversion efficiency and reduction of land space and reflector area to a minimum.

Prove the ability of the U.S. SOLAR POWER SUPPLY coupled with the LONG TERM ENERGY STORAGE after initial startup charging, when fully charged to produce 180 Kw$_E$ continuous for a period of thirty-one days without additional solar charging to demonstrate long term storage.

Prove the ability of the U.S. SOLAR POWER SUPPLY to operate at sufficient solar-electrical efficiency such that when extrapolated to large system, less than two tenths of one percent (.00125) of United States Area (less than one large county in Nevada or Utah), when fitted with the U.S. SOLAR POWER SUPPLY can provide power equal to all the electrical power now produced in the United States (DOE 2017).

Funding   70 Million

Time   Three Years

Request   Cooperative effort with Private Corporations, Academic Institutions, and Government including Government Labs Such As Palo-Alto, Sandia, INL. To Quickly Complete U.S. SOLAR POWER SUPPLY PILOT PROJECT

Frequently Asked Questions . . . How Big Is The Unit?

U.S. Solar Power Supply

Size of Installation Depends Upon Electrical Demand

A U.S. Solar Power Supply is sized to a given service area by ascertaining the service area's total annual demand and sizing the total integral of the U.S. Solar Power Supply annual output in total Kwhrs$_E$ to equal the service area's total Kwhrs$_E$ demand per year *plus thirty per cent* after Long Term Energy Storage losses.

Captured solar energy is added to storage (via condensing sodium) day by day while energy is extracted from storage to meet the instantaneous, hourly and daily demand. On those days when the energy deposited is greater than the energy withdrawn, the excess energy is saved for those days when demand is in excess of the energy collected.

Depending upon the particular location and seasonal demand cycle, either the summer surplus is stored for use during the winter or conversely, in areas dominated by higher summer demand for air-conditioning, the winter surplus is saved for the summer.

Seasonal storage losses for large City Sized Long Term Energy Storage Systems will be less than .03% for over six months storage duration even assuming an average outside temperature of 0°F.

See "U.S. Solar Power Supply Pilot Project
Solar-Electric Conversion Efficiency Summary" contained herein. See also "U.S. Solar Power Supply Solar-Electric Conversion Efficiency" a 19 Page detailed derivation page following. See also U.S. Solar Power Supply City Sized System page following.

What is the U.S. Solar Power Supply?

The U.S. Solar Power Supply is a unique solar-electrical conversion system that overcomes the two problems that have long prevented the use of solar energy for large scale nationwide power production:

First: U.S. SOLAR POWER SUPPLY is more efficient than other systems in converting solar energy into usable electrical energy.

Being more efficient gives the U.S. SOLAR POWER SUPPLY the advantage of being both smaller in size and less costly than other solar electrical systems that produce an equivalent amount of power.

Second: The U.S. SOLAR POWER SUPPLY solves the problem of long term energy storage.

The U.S. SOLAR POWER SUPPLY can store energy for many months with very small losses to produce an uninterrupted flow of power irrespective of daily, monthly, or even seasonal fluctuations in available solar energy. Enough energy is stored the U.S. SOLAR POWER SUPPLY that a backup conventional power plant is not required for nights, cloudy days, or even overcast seasons.

For a U.S. SOLAR POWER SUPPLY CITY SIZED SYSTEM
to power entire cities, the unit's LONG TERM THERMAL ENERGY STORAGE
will store energy collected during the summer months for use during the winter when there is less solar energy available. Energy loss from storage during that time period: less than 0.3%. Please see U.S. SOLAR POWER SUPPLY CITY SIZED SYSTEM With Six Months Thermal Energy Storage Duration: Summer To Winter At Continuous Average Demand Output" a derivation of efficiency and size required for a system for a city contained herein following.

For a U.S. SOLAR POWER SUPPLY NATIONAL SIZED SYSTEM providing power to significant sectors of the United States, the percentage storage loss will be even less than U.S. SOLAR POWER SUPPLY NATIONAL SIZED SYSTEM. Please see U.S. SOLAR POWER SUPPLY NATIONAL SIZED SYSTEM Sized To Meet 100% Of All United States Present Electrical Demand Equal To All The Power Now Produced From Coal, Nuclear, Natural Gas, Hydroelectric, Wind, Including All Private Power Producers" a derivation of efficiency and size required for a system for a city contained herein following.

Is U.S. Solar Power Supply energy dependable?

Because the U.S. Solar Power Supply replaces a conventional power plant, rather than merely reducing the fuel cost of a conventional power plant, the capital cost of the U.S. solar power supply can be written off against the savings of the capital cost of the conventional power plant.

The U.S. Solar Power Supply can stand alone.

How can the U.S. Solar Power Supply be used?

The U.S. Solar Power Supply was conceived to provide all the electricity used in the United States including all ground based transportation: Hence a true U.S. Solar Power Supply

that would power the entire United States.

However, U.S. Solar Power Supply units can be built to meet almost any size electrical demand: From massive to moderate. Units can be built to power military installations of any size above a certain level, (approximately 180 Kw_E continuous) or villages and subdivisions of thirty or more homes located in the solar rich American Southwest.

Arrays can provide power for universities, towns, factories, and remote installations not presently served by high tension power lines; or many other entities requiring power that wish to be energy independent.

During the next thirty years US demand for electricity is expected to double: More so if the electrical car should gain wide acceptance. Arrays of U.S. Solar Power Supply conversion and storage units, sited in the vast, virtually vacant American Southwest, coupled with a merged, in phase national transmission grid, can provide the power to meet the increase in electrical demand. Over the same next three decades or so, many old power plants will be phased out and the U.S. Solar Power Supply can provide their replacement power as well. The result, a large percentage of the United States, and ultimately Canada and Mexico too can be powered by The U.S. Solar Power Supply

A U.S. Solar Power Supply with Long Term Thermal Energy Storage - six months to a year or even longer - can be used for remote bases, and bomb proof bunkers. A U.S. Solar Power Supply can provide system that would allow military bases to be completely energy independent regardless of the time of year or the weather.

What are some of the financial consequences of a Long Term Thermal Energy Storage ?

Picture a synchronized in phase electrical power system that will provide electricity to as many electric vehicles as there are now conventional.

What are some of the environmental consequences of a U.S. Solar Power Supply?

A U.S. Solar Power Supply
driving a single in phase national power system coupled with universal use of electric vehicles would eliminate exhaust fumes, smoke, conventional coal mines, strip mines, radioactive byproducts, greenhouse gasses, and thermal pollution in rivers and lakes now serving as condenser coolants for our presently polluting power plants

What are some of the strategic consequences of a U.S. Solar Power Supply ?

The strategic implications of an energy independent United States are noteworthy. Would the elimination of imported oil from the Mid-East have positive National Security consequences and reduce the need for American overseas interventions? Possibly. Could we eliminate the need to secure the long shipping lanes?

What are some of the geopolitical consequences of a U.S. Solar Power Supply ?

There is an electric car that can go from zero to sixty in under four seconds. One could cruse over 300 miles before its battery module is quickly replaced at a switching station using light weight elements (i.e. An Aluminum-Air that are electrically rechargeable via the U.S. Solar Power Supply

How much land would it take for the U.S. SOLAR POWER SUPPLY?

If a fully realized U.S. SOLAR POWER SUPPLY were constructed to meet all the current electrical power requirements of the entire United States, along with all the projected increased electrical demand of the nation over the next thirty years, the land required for the collectors and storage facilities would be well less than one half of one percent of the nation's land area. Please see Papers Following: "U.S. SOLAR POWER SUPPLY CITY SIZED SYSTEM With Six Months Thermal Storage Duration - Summer Through Winter At Continuous Average Demand Output", herein page 16, and see U.S. U.S. SOLAR POWER SUPPLY NATIONAL SIZED SYSTEM With Six Months Thermal Storage Duration – Summer Through Winter At Continuous Average Demand Output" also herein following.

U.S. SOLAR POWER SUPPLY PILOT PROJECT

Areas of Technical Interest And Milestones

Such U.S. SOLAR POWER SUPPLY PILOT PROJECT will demonstrate the efficacy of the technology of the U.S. SOLAR POWER SUPPLY in the following:

Efficiently Intercept, Concentrate And Capture Solar Energy In Solar Boiler

Vaporize Sodium in Solar Boiler

Condense Sodium via Heat Exchanger Immersed In NaCl Tank
To Deposit Thermal Energy In Molten Salt

Demonstrate LONG TERM THERMAL ENERGY STORAGE
- more than 30 days - at less than 2% loss

Produce super critical pressure steam upon demand at 1350°F from storage of sufficient quantity to :

Produce 620KwE solar/electrical power at 36.807 % overall efficiency

Produce from storage 180KwE continuous for more than 30 days without sun

Produce 1 Megawatt Peak Power

Phase II (see following) of the U.S. SOLAR POWER SUPPLY PILOT PROJECT will consist of overseeing the detailed design and its reduction to final construction drawings accompanied by fabrication and assembly instructions for the first unit.

Main areas of concern are:

Sodium Boiler Design

Sodium Condenser inside LONG TERM THERMAL ENERGY STORAGE Design

Molten Salt Container and Insulation Design

Foundation, Main Support, Reflector Support, and Boiler Support Design

Steam Heat Exchanger at Super Critical Pressure with Reheat

Areas of Technical Interest And Milestones

Other technical areas of interest include designing and/or implementing:

Standards of Practice - Materials Handling - Namely
    Sodium Vapor
    Molten NaCl
    Super Critical 1350°F. Steam

Sodium Pump, Counterflow Fluid / Vapor Line
Rotating Sodium / Steam Lines (One Rotation Per Day)

Design 1 Megawatt Super Critical 1350 °F. Steam Turbine Throttleable Down To 50Kw$_E$
Regenerative Feedwater Heater Design
Air Cooled Exhaust Steam Condenser and Feed Water Pumps
Solar Tracking
Reflector Element Support Design
Reflector Superstructure Design
Controls And Data Collection

Phase I of the LONG TERM THERMAL ENERGY STORAGE lasting 4 months: Set up local office, hire two support staff, administrative-financial, internal IT support, contact and meet with individuals in areas of expertise as outlined above, travel included as required. Projected Cost $1,750,000

Phase II of the U.S. SOLAR POWER SUPPLY PILOT PROJECT lasting six months: Utilize expertise from several research universities such as MIT, Virginia Tech, Stanford, Government Labs, such as INL formally Argonne West having over thirty years' experience with sodium, locate sites for building and testing individual components parts, locate and lease or buy site for construction of first unit.. Projected Cost $8.75 Million.

Phase III of the U.S. SOLAR POWER SUPPLY PILOT PROJECT lasting a year, will involve the actual, constructing and testing the various components. Projected Cost $17.5 Million.

Phase IV of the U.S. SOLAR POWER SUPPLY PILOT PROJECT lasting eight months will be the construction of the first prototype of the U.S. SOLAR POWER SUPPLY including procurement of the assembly building, materials, ready-made components, tools, and man power needed for the final erection and operation. Projected Cost $35 Million.

Phase V of the U.S. SOLAR POWER SUPPLY PILOT PROJECT lasting one year will be the data collection and cyclical solar seasonal operation of the unit. Projected Cost $7 Million.

U.S. Solar Power Supply

World's Most Efficient Solar Electrical Power Production With Long Term Thermal Energy Storage Requires Least Land Area

U.S. Solar Power Supply Pilot Project

Solar - Electric Conversion Efficiency: *Summary*

45 Meter Tracking Reflector Dish
With Thirty-One Days Storage at Average Continuous Rated Output

Solar Energy Available

Area Of Dish = 1590.43128 M^2

Direct Solar Energy = 1.044 Kw$_{\text{direct solar}}$ / M^2 = 1.044 Kw$_s$ /M^2

Solar Energy Intercepted = 1590.43128 M^2 x 1.044 Kw$_s$ / M^2

= 1660.41025 Kw$_s$

Solar Energy To Reflector After Boiler & Boiler Support Shadows

1660.41025 Kw$_s$ - 1% Shadows = 1643.8061 Kw$_s$ To Reflector

Solar Energy To Reflector Elements After Gaps Between Elements

1643.8061 Kw$_s$ - 1/4% Gaps Area = 1639.6965 Kw$_s$ To Elements

Solar Energy To Reflector Element Surface After Dust

1639.6965 Kw$_s$ - 1/4% Area Dust Factor = 1635.5972 Kw$_s$ To Surface

Solar Energy To Undamaged Element Surface After Abrasion

1635.5972 Kw$_s$ - 1/4% Abrasion Factor = 1631.5082 Kw$_s$ To Intact Surface

Solar Energy Reflected Toward Solar Boiler

1631.5082 Kw$_s$ To Intact Surface x
88% Secular Reflectivity Alcoa R5 Aluminum = 1435.7272 Kw$_s$ Toward Boiler

Solar Energy Entering Boiler Aperture

1435.7272 Kw_s x 99.5% Through Boiler Aperture = 1428.5485 Kw_s Into Boiler

Solar Energy Absorbed In Boiler After Most Rays Subjected To Three Internal Reflections

1428.5485 Kw_s x 99% Absorbed = 1414.263 $Kw_{ThermalA}$ Absorbed Thermal Energy

Retained In Boiler After Heat Losses

1414.26 Kw_T - 96.00 Kw_T Radiation
                     - 11.13 Kw_T Convection
                     - 4.01 Kw_T Conduction = 1303.12 Kw_T In Sodium

Thermal Energy Delivered To Thermal Storage after Sodium Line Heat Loss
1303.12 Kw_T - 6.88 Kw_T Sodium Line Loss = 1296.24 Kw_T To Storage

Net Thermal Energy Extracted After 31 Days In Thermal Storage
1296.24 Kw_T - 1.576% Loss After 31 Days = 1275.82 Kw_T Extracted

Electrical Energy Produced
1275.82 Kw_T x 48.619% Thermal-Electrical Efficiency Turbine-Generator***

= 620.29 Kw_E (Solar Electricity Produced)

Sodium Pump Loss
      620.29 Kw_E - 1.87 Kw_{ESP} = 618.42 Kw_E After Sodium Pump

Solar Clock Drive Loss
618.42 Kw_E After Sodium Pump - 6.59 $Kw_{SolarClockDriveEquivalent}$ =
= 611.83 Kw_E After Clock Drive

Lighting
611.83 Kw_E After Clock Drive - .68 $Kw_{LightingEquivalent}$ =

= 611.15 Kw_E Net After All Parasitic Losses

Solar Electrical Conversion Efficiency

    611.15 Kw_E = 36.807% Solar To Electrical Efficiency***

1660.41 Kw_{SD}

Continuous Output From Storage

7.07 Hours / Day Of Full Rated Sun Equivalent x 611.15 Kw_E

= 4320.83 $Kwhr_E$ / Day

$$\frac{4320.83 \; Kwhr_E \; / \; Day}{24 \; hours \; / \; Day} = 180 \; Kw_E \quad \text{Continuous}$$

***Please See U.S. SOLAR POWER SUPPLY Solar-Electric Conversion Efficiency" A 19 Page Detailed Derivation contained herein following

U.S. SOLAR POWER SUPPLY PILOT PROJECT

Solar Electric Conversion Efficiency

Detailed Derivation

45 Meter Diameter Solar Normal Tracking
Reflecting Paraboloidal Dish with Thirty-One
Days Storage at Average Continuous Rated Output

The Direct Solar Flux

The direct solar flux at sea level is often 900 watts per square meter. At 5500 feet elevation the direct solar flux is 16% more or 1044 watts per square meter, and at Fort Huachuca, Arizona the direct solar flux is greater. For this purpose a value of 1044 watts per square meter will be used.

Reflector Dish

The reflector dish will be a 60° Paraboloidal. The diameter of the reflector dish will be 45 meters. Normal reflecting surface area is PI x (1/2 x 45 Meters)^2 = 1590.43128 Meters^2

Direct Solar Flux To Gross Reflector Area

1.044 KwE_{SD} / M^2 x 1590.43128 Meters^2 = 1660.41025 Kw$_{SD}$

Solar Flux To Reflector Area After Boiler And Boiler Support Shadows

Boiler aperture area derived herein following at 1.3362 Meters^2
Boiler supports shadow designed at 14.5683 Meters^2 (eight very narrow flange deep webbed I beam supports)

1660.41025 Kw$_{SD}$ – 1% Total Shadows = 1643.8061 Kw$_{SD}$

Solar Flux To Reflector After Gaps Between Elements

1643.8061 Kw$_{SD}$ – ¼ % Gaps Area =
=1639.6965 Kw$_{SD}$ To Elements Surface

Solar flux To Reflector Surface After Dust

1639.6965 Kw$_{SD}$ – ¼ % Area Dust Obstruction = 1635.5972 Kw$_{SD}$

Solar Flux To Undamaged Reflector Element After Abrasion

1635.5972 Kw$_{SD}$ - ¼ % Abrasion Obstruction = 1631.5082 Kw$_{SD}$

Individual Triangular Reflecting Elements

The reflector dish will be composed of 297 triangular reflecting elements arranged to form a paraboloidal.

Reflectivity Of The Reflector Surface

Freshly evaporated aluminum on a smooth surface has a total normal (specular) reflectivity for solar radiation of 92%. Alcoa R5 Bright Dip treatment has an 88% specular reflectivity.

Solar Energy Reflected Specular-ly (At Equal Angle To Incident)

1631.5082 Kw_{SD} x 88% Secular Reflection = 1435.7272 Kw_{SD} Toward Boiler

The width of each element at its base, the edge of the reflector, line AB, is (PI x 45 meters) / 297 elements = .4759 meters

Boiler To Reflector Distance

Maximum boiler – reflector distance for a 45 meter 60° paraboloidal is 31.26 meters. The solar half angle is 16 minutes, the sun's angular width is 32 minutes.

Size Of Solar Image At Boiler

Ideal image width at boiler is .4749 meters base length + (2 x sin 16) x 31.26 Meters = .766 meters,

Radius of Ideal Image is .383 Meters

Vertical Support Tilt Angle

Specify two reflector element vertical end supports AA and BB are cut to length within 0.2 mm tolerance.

297 Individual Reflecting Triangular Elements Forms Paraboloidial Reflector

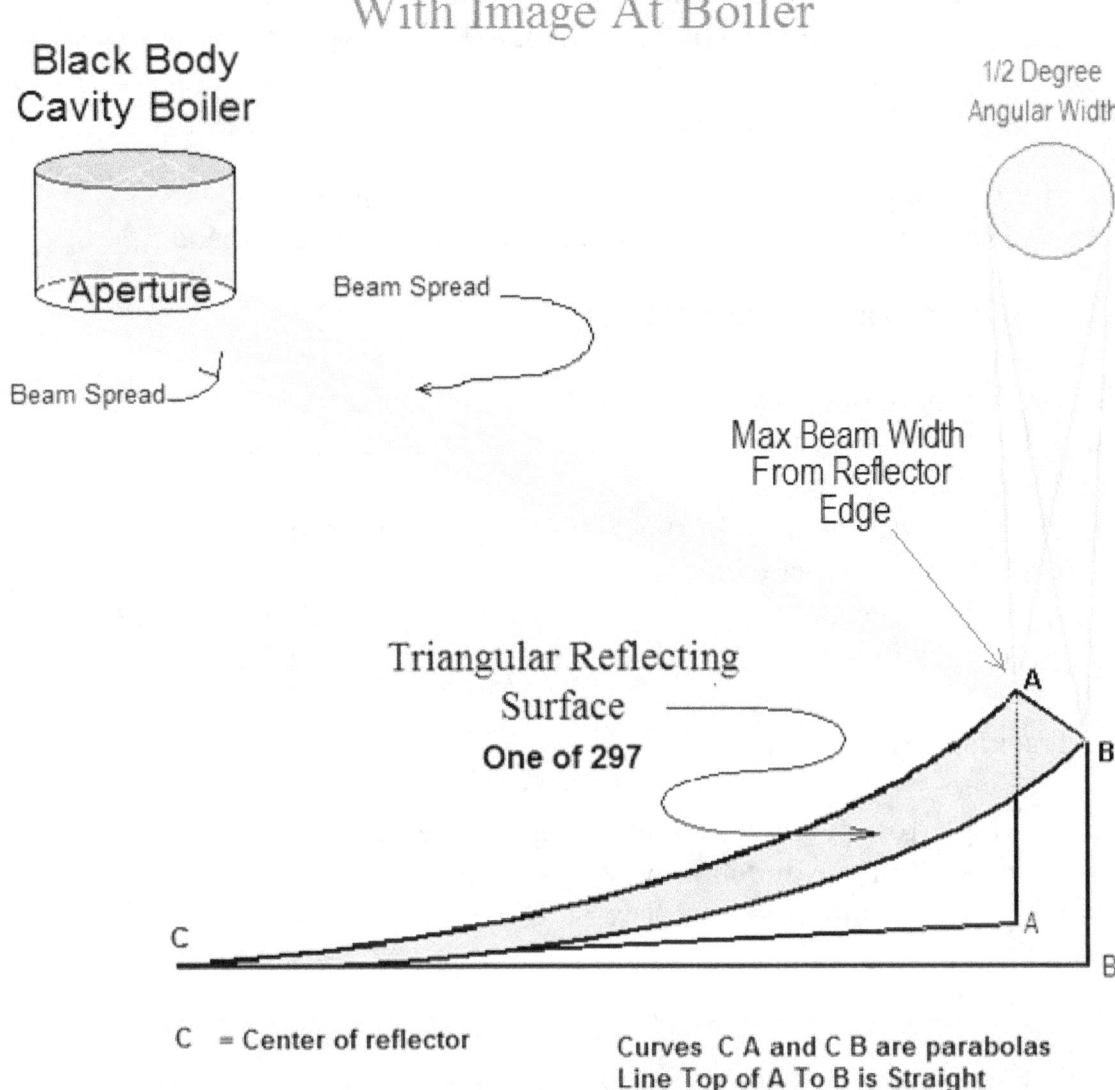

Let support AA be .2mm too short, while support BB be .2mm too long to produce .4mm variation. Sin of maximum reflector element tilt angle, $RETA_{VSE}$ is .4mm / 475.9mm base length AB of focal image due to vertical support error, $Shift_{VSE}$ = .4mm / 475.9mm x 31.27 meters = .02627 meter

Lateral Support Pipe Diameter Deviation Tilt Error

Alcoa Standard Construction Tube deviation of pipe diameter is .015" = .381mm

Let curved support pipe CA be maximum deviation thick while curved support pipe CB be maximum deviation thin.

Maximum variance between both lateral supports is .761 mm

Sin of Maximum Reflector Element Tilt Angle $RETA_{LSPDE}$ at base =
= .762mm / 475.9mm

$Shift_{LSPDE}$ of focal image = .762 / 475.9 x 31.26 Meters = .05 meter

Support Pipe Non-Conformity To Parabola +/- .5 Degree
Increases Depth of image but not width; Boiler area does not expand

Undercarriage Support Droop Tilt Error

Reflector support along an axis is subject to droop along that axis. Assume that center of reflector droops 0.1 meter from specifications at center.

There are 74.25 individual reflecting elements per quarter reflector, there is 1.35mm droop per element

Sin of maximum element tilt angle at base $RETA_{LSPD}$ is 1.35mm / 475.9mm

$Shift_{USDT}$ of focal image = 1.35 / 475.9 x 31.26 meters = .0887 meters

Solar Tracking Error

Solar tracking error specified at or less than 1/9 degree.
Focal image shift in one direction only
Focal image shift, $Shift_{STE}$ = sin 1/9° x 31.26meters = .0606meters

Wind Load Error

Specify deformation at reflector axis perpendicular to seasonal support gimbals due to average assumed 9 mph wind is .05 meter
Calculated in the same way as the droop error
Focal image shift, $Shift_{WLE}$ = .0436meters

Sum Of Focal Errors

Virtually impossible worst case where all errors act on each individual reflector element in the same direction to produce the largest possible total shift on that element

$Shift_{Total} = Shift_{VSE} + Shift_{LSPDE} + Shift_{USDT} + Shift_{STE} + Shift_{WLE} =$

$= 02627M + .05M + .0887M + .0909M + .0436M = .2691M$

Actual Image Area

Even more impossible worst case where not only do all errors act on each individual element in the same direction, but where they act on each adjacent individual reflecting element in the opposite direction in such a manner as to produce the largest possible image area at the boiler aperture.

$PI \times (Ideal\ Radius + Shift_{Total})^2 =$

$= PI \times (.383 Meters + .26917)^2 = 1.3362 Meters^2$

Boiler Aperture Area

The boiler aperture area is set at 1.3362 Meters²

Solar Energy Entering Boiler Aperture

To allow for additional unforeseen reflector deviations or tracking error only 99.5% of reflected direct solar flux will be counted as passing through the boiler aperture.

1435.7272 Kw_{SD} Toward Boiler × 99.5% = 1428.5485 Kw_{SD} entering boiler aperture.

Flux Absorption In The Boiler

Stainless steel exposed to high flux density exhibits a solar absorption of .8

Proper ceramic optical coating with craggy surface smoothed to eliminate only top pointed surface yet retain very high fissure ratio can yield a spectral reflecting absorber with .9 solar absorption. After two reflections in the boiler :

100% entering flux x 90% absorption 1st reflection = 90% absorbed and 10% reflected x 90% absorbed 2ed reflection = 9% more absorbed 2ed reflection = 99% absorbed after two reflections.

Three fourths of all rays from the reflector are subject to three reflections in the boiler for a total 99.675% absorbed.

We will assume that only 99% of incoming flux is absorbed.

1428.5485 Kw_{SD} entering boiler aperture x 99% = 1414.263 Kw_{TEA}
Thermal Energy Absorbed (TEA) at ceramic surface

Energy Across Ceramic Absorber Boiler Surface

Energy Absorbed = 1414.263 Kw_{TEA} = 4,826,879.7 Btu/Hr_{TEA}

Ceramic coating thickness is .0168In

Conductivity of ceramic = K_C = 5Btu/HrFt² °F/Ft

5Btu/HrFt² °F/Ft / 12 inches / Ft = 60Btu/HrFt² °F/inch

60Btu/HrFt² °F/inch / .0168" Actual Thickness = 3571Btu / HrFt² °F

Let inside height, H_B, of cavity boiler equal the width of boiler aperture

Boiler aperture area = PI x (R_{BA})²

R = square root of (boiler aperture area 1.3362 Meters² / PI) =
=.6522 Meters

Height Inside Boiler = H_B = 2 R =
=1.3043 Meters, total surface area of inside (closed on top cylinder) = PI x R² + H_B x PI x 2R = 6.68Meters² = 71.89585ft²

$3571 \text{ Btu/HrFt}^2 \text{ °F} \times 71.89585 \text{ ft}^2 = 256{,}740 \text{ Btu/Hr°F}$

$4{,}766{,}979.739 \text{ Btu/Hr}_{TEA} / 256{,}740 \text{ Btu/Hr°F} = 18.85 \text{°F}$ Delta T Across Ceramic

Energy Across Tube Walls In Boiler

To form inner surface without normal surface presented to incoming solar rays, tubes are square channel carrying sodium liquid > vapor at slightly above atmospheric, Wall thickness = .1214"

Tube wall surface is four times boilers absorbing surface, however due to relative resistance of barrier layer of sodium on inside surface verses tube wall conductivity, only half of tube's surface area will be used a heat conducting surface to sodium

Area of tube surface = $4 \times 71.89585 = 287.5834 \text{ Ft}^2 / 2 = 143.7917 \text{ Ft}^2$
Conductivity of stainless steel tube $K_P = 10 \text{ Btu/HrFt}^2 \text{ °F/Ft} =$
$= 120 \text{ Btu/HrFt}^2 \text{ °F/Inch}$
$120 \text{ Btu/HrFt}^2 \text{ °F/Inch} / .1214"$ Thickness $= 988.4678 \text{ Btu/HrFt}^2 \text{ °F}$
$988.4678 \text{ Btu/HrFt}^2 \text{ °F} \times 143.7917 \text{ Ft}^2 = 142{,}133.46 \text{ Btu/Hr°F}$
$4{,}826{,}879.7 \text{ Btu/Hr Btu/Hr}_{TEA} / 142{,}133.46 \text{ Btu/HrFt}^2 \text{ °F} =$
$= 33.96$ Delta T °F across pipe wall

Energy Across Sodium Boundary Layer
Conductivity of sodium is 31 Btu/HrFt^2 °F

Let the boundary layer be .05 inches thick (inversely proportional to: mean vapor velocity, inside pipe diameter, and solar flux / thermostatically controlled mass flow)

$31 \text{ Btu/HrFt}^2 \text{ °F/Ft} / 12 \text{ inches/Ft} = 372 \text{ Btu/HrFt}^2 \text{ °F/inch}$
$372 \text{ Btu/HrFt}^2 \text{ °F/inch} / .05"$ Actual Thickness $=$
$= 7440 \text{ Btu/HrFt}^2 \text{ °Fs}$
$7440 \text{ Btu/HrFt}^2 \text{ °F} \times 143.79 \text{ Ft}^2 = 1{,}069{,}797.6 \text{ Btu/Hr °F}$
$4{,}826{,}879.7 \text{ Btu/Hr}_{TEA} / 1{,}069{,}797.6 \text{ Btu/Hr °F} = 4.512$ Delta T °F sodium boundary layer

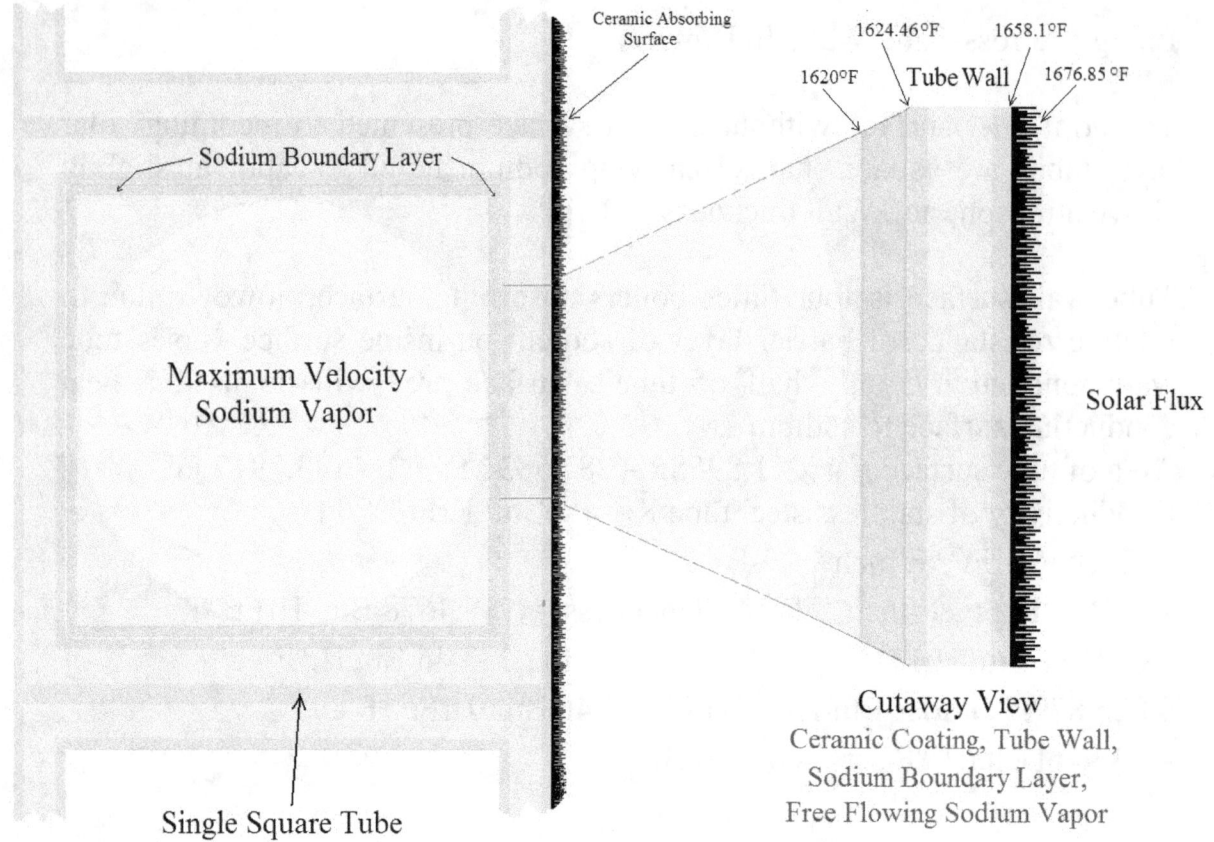

Temperature Of The Boiler Surface

By thermostatic control, the sodium vapor will be kept above 1620°F at the header where the vapor is channeled into many pipes immersed in common salt, NaCl, inside the LONG TERM THERMAL ENERGY STORAGE tank. Some degree of superheat at the boiler will be required to make up for energy required for regenerative pre heating of the incoming liquid condensed at 1620°F inside the tank and subsequently subcooled by the salt at 1486.4°F in its end steady state.

For purposes of *over* calculating heat loss from the solar boiler we will assume that the surface temperature of the boiler surface is not 1677.32°F as derived above, but 1683°F to yield a greater energy loss from the boiler. This five degrees of superheat will more than make up for heating slightly subcooled sodium feed resulting in isothermal vaporization in boiler.

Radiation Loss From Boiler

We will assume that cold sink temperature is absolute zero (worst case), boiler sees blue sky.
Boiler aperture radiates as a black body cavity at 1683°F = 1189.6°K.
R_o, the radiation coefficient is 5.67×10^{-8} watts / M^2
Selective surface ceramic emissivity at given temperature, e = .63
Area of aperture from above is 1.3362 M^2
Radiation Heat Loss = R_o x Area x T (°K)^4 = 95,571 watts   Round up to 96 Kw

Convection Loss From Boiler

Convection
Where Convection coefficient = C_{cc}
= C_{cc} x (Delta T °C)$^{1.25}$ x Area x sec/hr x Kwhr / 860,030 cal
Maximum case vertical surface: Convection coefficient C_{cc} =
= .4 x 10^{-4} cal / sec.cm^2
1683°F Boiler - 60°F Average Air =
Delta T = 1623 °F = 902°C Delta T
Area = 13,362cm^2 3600sec. / hr   860,030 calories / Kwhr
.4 x 10^{-4} cal / sec.cm^2 x (902°C Delta T)$^{1.25}$ x 13,362cm^2 x 3600sec/hr x Kwhr/860,030calories = 11.13 Kw (Kwhr/hr) Maximum convection Loss When Boiler Aperture Is Vertical (6 AM or 6 PM) Boiler operates at isothermal temperature of 1683°F

Conduction Loss From Boiler

Boiler walls are insulated with two layers of insulation.
Inner high temperature insulation conductivity K_1 = .1 Btu / hrft°F
Outer lower temperature insulation conductivity K_2 = .05 Btu / hrft°F
T_{hot} = not above 1625°F (cool side of outer side of boiler tube wall)
T_{cold} = 24°F, Ln = natural log
$H_{cylinder}$ = 4.2788ft, $D_{aperture}$ = 4.2788ft, R_1 = 2.1394ft, R_2 = 3.3247ft, R_3 = 4.433ft

Conduction $_{Cyl\ Walls}$

$$\frac{PI \times D \times L \times (T_{hot} - T_{cold})}{(Ln(R_2/R_1)/K_1) + (Ln(R_3/R_2)/K_2)} = 9149\ Btu/hr$$

Conduction $_{Cir\ Top}$ =

$$\frac{PI \times \tfrac{1}{2}D \times \tfrac{1}{2}D \times (T_{hot} - T_{cold})}{(\tfrac{1}{4}D/K_1) + (\tfrac{1}{4}D/K_2)} = 698\ Btu/hr$$

Additional Shape Factor conduction losses

Conduction bottom edge = 1/4 walls = 2287 Btu/hr

Conduction top corner = 1/6 walls = 1524 Btu/hr

Conduction total = 13,658 Btu/hr = 4.002 Kw

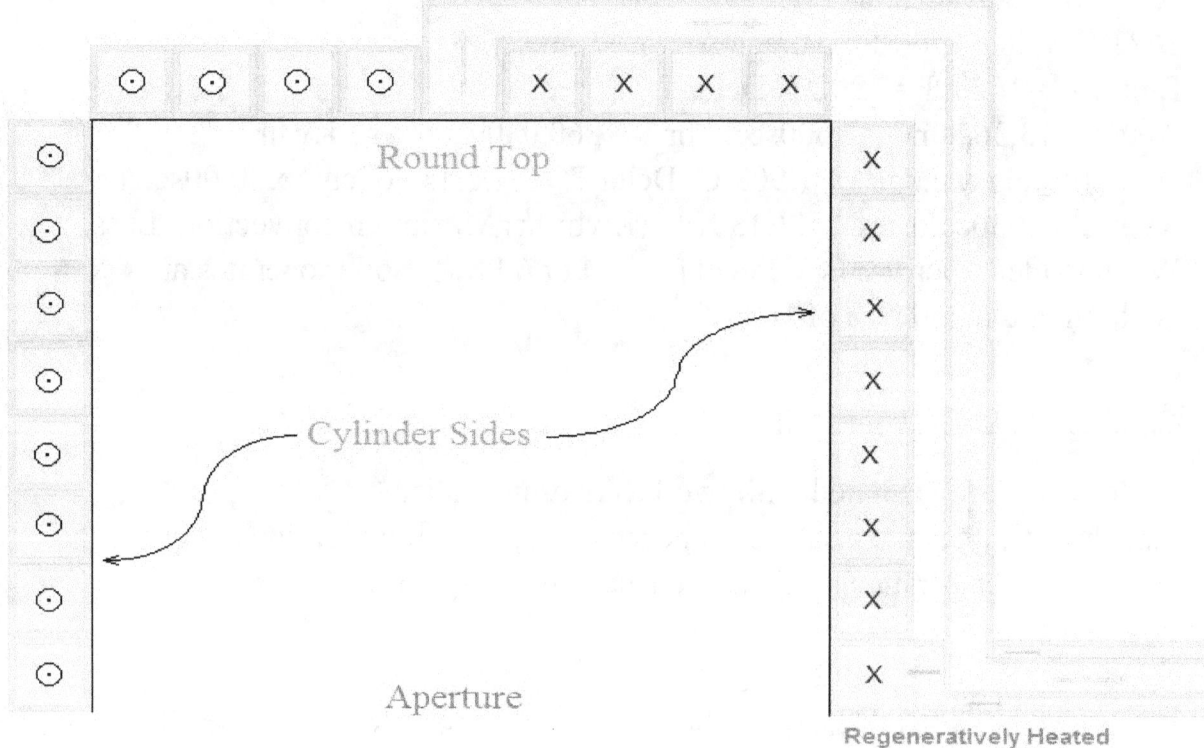

Schematic Of Black Body Solar Boiler
Round Top and Cylinder Side Lined
With Square Channel Tube

Heat Loss Sodium Counterflow Line = $H_{\text{sodium line}}$

Inside sodium vapor line OD = 4", Outer sodium fluid line OD = 4.8" = .4ft

Length from boiler to insulated tank = 260ft

T_{hot} = 1640°F Superheated sodium vapor as required for regenerative heating

T_{cold} = 0°F outside air temp

2 layers of insulation, K_1 = .1 Btu / hrft°F, K_2 = .05 Btu / hrft°F

R_1 = .2ft, R_2 = 1ft, R_3 = 1.4ft

$H_{\text{sodium line}}$ =

$$\frac{\text{PI} \times D \times L \times (T_{\text{hot}} - T_{\text{cold}})}{(\text{Ln}(R_2/R_1)/K_1) + (\text{Ln}(R_3/R_2)/K_2)}$$

= 23,491 Btu/hr = 6.88 Kw

```
1414.263 KwTEA  thermal energy absorbed at ceramic surface
 -96.57 KwRad
 - 11.13 KwConvSB
 -  4.01 KwCondSB
 -  6.88 KwSodium Line
1296.25 KwNetToThermalStorage
```

Thirty-One Day Thermal Storage For 45 Meter U.S. SOLAR POWER SUPPLY PILOT PROJECT

Size of U.S. SOLAR POWER SUPPLY is designed to meet average integral of electrical demand over entire year. For a northern temperate area with cool to cold winter, the solar energy deficit from September to March is somewhat less than thirty days average output. Thus a large U.S. SOLAR POWER SUPPLY CITY SIZED SYSTEM in order to provide continuous un-supplemented power through the winter until the spring would need to be able to efficiently store enough thermal energy for thirty day's output (net no additional solar charging).

Maximum Electrical Need:

7.4 kwhr/M² day direct solar flux / 1.044Kw/M² = 7.07hrs/day average solar input

7.07 hrs/day x 620.2909 Kw$_{ElectricInternalProduced}$/ 45 meter unit *** = =4385.4566 Kwhr$_E$ / 24 hr day

4385.4566 Kwhr $_E$ / 24 hr day = 182.72 Kw$_E$ Continuous Output

Monthly Need Of Energy To Be Stored
182.72 Kw$_E$ x 24hrs/day x 31 days/month = 135,943.68 Kwhr$_E$
135,943.68 Kwhr$_E$ / 48.619% Thermal-Electrical efficiency*** of turbine-generator =
= 279,610.19 Kwhr$_T$

Volume Of Molten Salt

Heat of Fusion Molten Salt = 8.286 KwHr$_T$/Ft³

= 279,610.19 Kwhr$_T$ / 8.286 KwHr$_T$/ft³ = 33,744.89 ft³ For 31 days thermal storage

***See Thermal Electrical Derivation Following

Size Of Cube To Contain Thermal Energy Needed
Cube root of 34,000 ft³ = 32.3961 feet
Round Up. Thermal Storage is a cube 34 on an edge.

Energy Transfer To Thermal Storage Assuming No Convection In Molten Salt

1296.25 Kw$_{NetToThermalStorage}$ = net energy in sodium vapor after all reflector, boiler and sodium line losses

1296.25 Kw$_{NTTS}$ x 3413 Btu / Kwhr = 4,424,101.3 Btu/hr required heat transfer

While pipes with fins can be used to reduce size and length of pipes needed, for simplicity, only straight conduction through solid salt from bare pipes will be presented.

Let sodium vapor be divided at a heat transfer headers into 1444 separate 1-1/2 " schedule 40 pipes 34 feet long spaced evenly in tank (38 pipes x 38 pipes)

Diameter of 1 ½" pipe = .15832 feet   Radius = .07916 feet
Volume of 1 ½" pipes in cubic feet
PI x (.07916 ft)^2 x 34 ft length x 1444 pipes = 966.51 ft^3 for the heat in pipes
Volume of 1" Heat Out Pipes (From Below)
PI x (.0234Ft)^2 x 34 ft length x 1000 pipes = 58.49 ft^3 for the heat out pipes

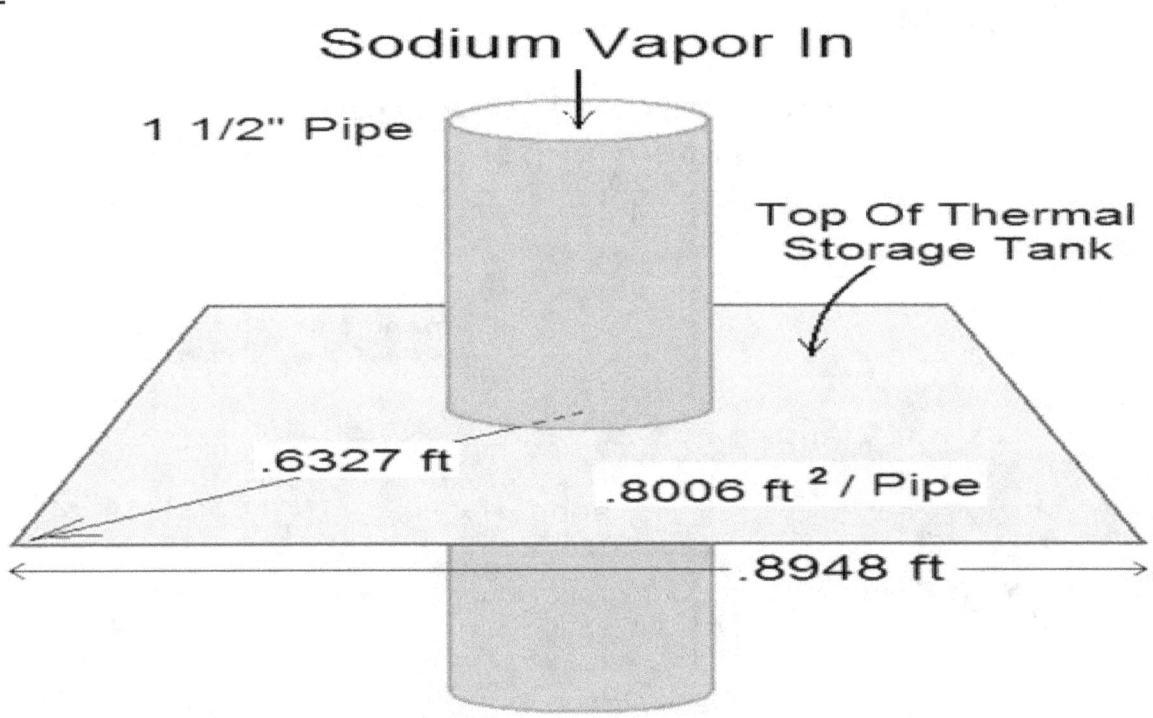

Area Of Thermal Storage Tank Top Per Heat In Pipe

(34ft x 34ft) / 1444 pipes = .8006 ft² / pipe

So each pipe is going into center of an area of = .8006 ft²

Inner radius of pipe = .06708 = R_1, Outer radius of pipe = .07916 = R_2

Maximum radius (heat flow distance) to last vestige of molten salt at full charge : (~Sept 21, normal steady state operations)

Square root of (2 x (.8006 ft / 2)² = .6327 ft ~ .633 ft = R_3

K_1 of pipe = 8.58 Btu / hrft°F, K_2 of salt = 2.8037 Btu / hrft°F, Ln = natural log

Sodium temperature = 1620°F, Salt temperature = 1486°F

Energy Transferred =

$$\frac{2 \times R_2 \times PI \times 34ft \times 1444 \times (1620 - 1486)}{[(Ln(R_2/R_1)/K_1) + (Ln(R_3/R_2)/K_2)]} = 4{,}493{,}546.3 \text{ Btu}$$

4,493,546.3 Btu > 4,424,101.3 Btu/hr required heat transfer for last day of thermal charging when heat has to travel greatest distance to melt remaining solid salt at perimeter

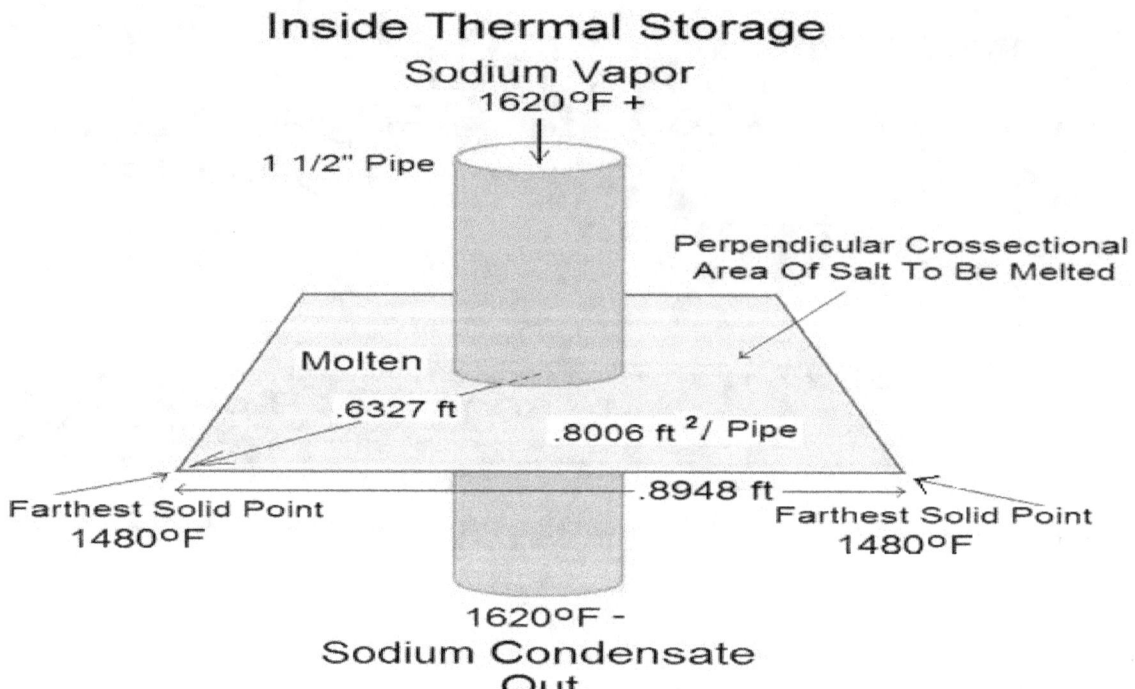

Energy Extraction From Thermal Storage

Continuous Rated Electrical Output = 182.94810 Kw$_E$ ~ 182.9   Kw$_E$

180.24 Kw$_E$ / .48619 Thermal Electrical Efficiency = 370.71 Kw$_T$ = 1,265,233.2 Btu/hr

This energy must be able to be withdrawn on the *last* day of drawdown from thermal storage
(~ March 21 during normal steady state operations)

While pipes with fins can be used to reduce size and length of pipes needed, for simplicity, only straight conduction through molten salt to bare pipes will be presented.

Regeneratively preheated high pressure feed water enters thermal storage at 370 °F , is heated and exits at 1350 °F

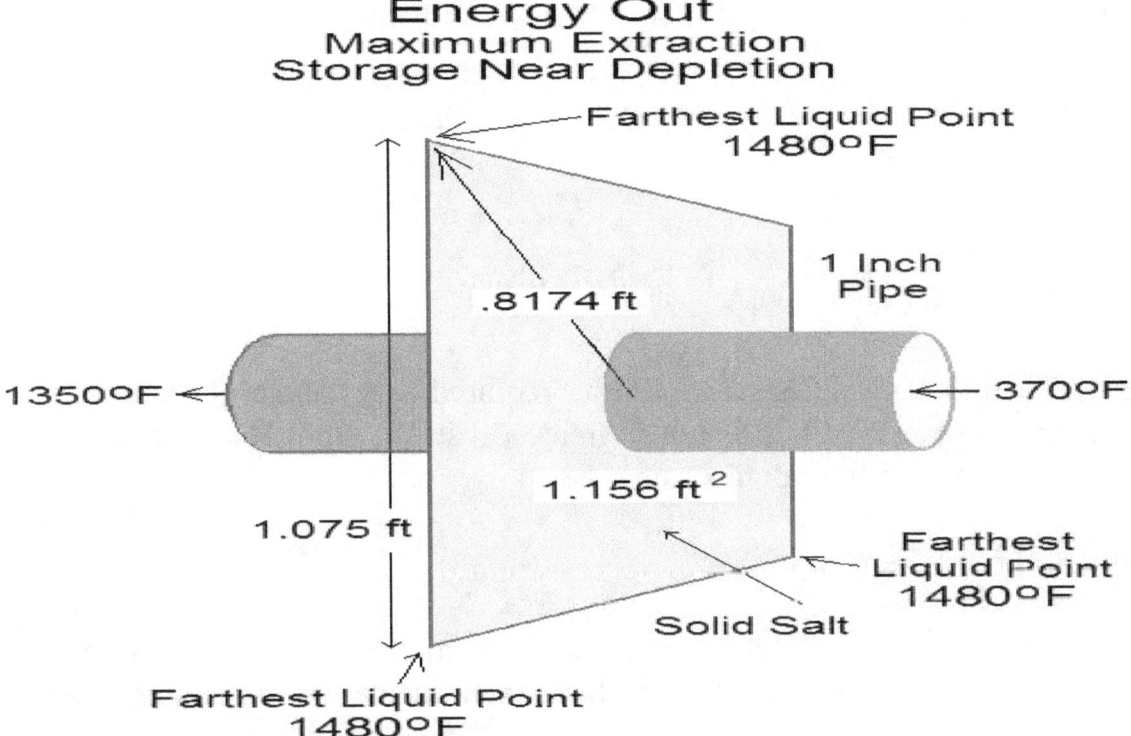

The log mean temperature difference between salt and water (LMTD)

$$\frac{(1480-370) - (1480-1350)}{Ln(1480-370) - Ln(1480-1350)} = 446.34 \text{ LMTD}$$

1000 Pipes 1" Schedule 80, Each pipe is 34 feet long,

Each heat out pipe sits in an individual square of 1.156 ft² which is 1.075 ft on a side

Maximum distance (at maximum energy draw down) for peak power production when energy moves from last vestige molten salt through solid salt at full discharge is = R_3

Square root of $(2 \times (1/2 \times 1.156 \text{ ft}^2)^2 = .8174 \text{ ft} = R_3$

Inner radius of pipe = .0125 ft = R_1,

Outer radius of pipe = .0234 = R_2

K_1 of pipe = 8.58 Btu / hrft°F,

K_2 of salt = 2.8037 Btu / hrft°F , Ln = natural log

Energy Transferred =

$$\frac{2 \times PI \times 1000 \times 34 \times 446.34 \, (LMTD)}{[(Ln(R_2/R_1)/K_1) + (Ln(R_3/R_2)/K_2)]} = 1{,}343{,}078 \text{ Btu/hr}$$

1,343,078 Btu/hr > 1,265,233.2 Btu/hr required heat transfer for last day of drawdown when heat has to travel greatest distance from last vestige of molten salt through solid salt to steam pipes.

Heat Loss From Thermal Storage

Thermal storage tank will be a cube 34 feet on an edge covered with three layers of insulation. Pilot project will have thick insulation to demonstrate high efficiency. Units twice as big, (edge equal to 68 feet) will hold eight times the energy but will suffer only four times the loss.

Layer I_1 = 3 Feet, Layer I_2 = 4 Feet, Layer I_3 = 5 Feet, Layer I_4 = 3 Feet,

Layer K_1 = .1 Btu/hrft°F, Layer K_2 = .05 Btu/hrft°F, Layer K_3 = .015 Btu/hrft°F,
Layer K_4 = .01 Btu/hrft°F,

Assume ambient air temp 0°F for 31 Days. $T_{hot} - T_{cold}$ = 1480°F - 0°F = 1480

Heat loss walls =

$$\frac{6 \text{ Walls} \times 34\text{ft} \times 34\text{ft} \times T_{hot} - T_{cold}}{I_1/K_1 + I_2/K_2 + I_3/K_3 + I_4/K_4} =$$

=13,336.52 Btu/hr

Heat losses edges and corners add 51.8% using standard shape factor

1.518 x 13,337 Btu/hr = 20,245 Btu/hr

Thermal loss 31 days = 20,2745Btu/hr x 24hr/day x 31 days = 15,062,280 Btu

15,062,280 Btu / 3413 Btu/Kwhr = 4413.21 Kw_T

279,959.2472 $Kwhr_{T \text{ Stored}}$ Total energy stored in thermal

4413.21 $Kw_{T \text{ Lost}}$ / 279,959.2472 $Kwhr_{T \text{ Stored}}$ = 1.576% loss

1296.25 $Kw_{NetToThermalStorage}$ - 1.576% loss from thermal storage** =

=1275.82$Kw_{Net From ThermalStorage}$

**This loss of 1.576% is only for very small increment of energy stored for maximum storage time of 31 days. Under normal steady state usage, 70%

of energy deposited is retrieved within one day, 25% within one week, 4% within two weeks.

Net recovered from thermal storage 275,541.44 Kwhr$_{T\ Recovered}$ =

$\dfrac{275,815.68\ \text{Kwhr}_{T\ Recovered}}{(24\ \text{hr}\ \times\ 31\ \text{days})}$ = $\dfrac{370.72\ \text{Kwhr}_T}{\text{Per ave hr}}$

370.72 Kwhr$_T$ x 48.619% thermal to electrical efficiency (see below)

= 180.24 Kw$_E$ Per ave hr

Turbine Efficiency (used in above)

Turbine cycle description: Two stage turbine with one high and one low pressure section

Two steam extractions for two regenerative feed water heaters

Initial steam condition from thermal storage: 7000psi at 1350°F

Reheat in thermal storage to 1650psi at 1350°F

Exit steam condition to condenser: .25psi at 60°F (Average)

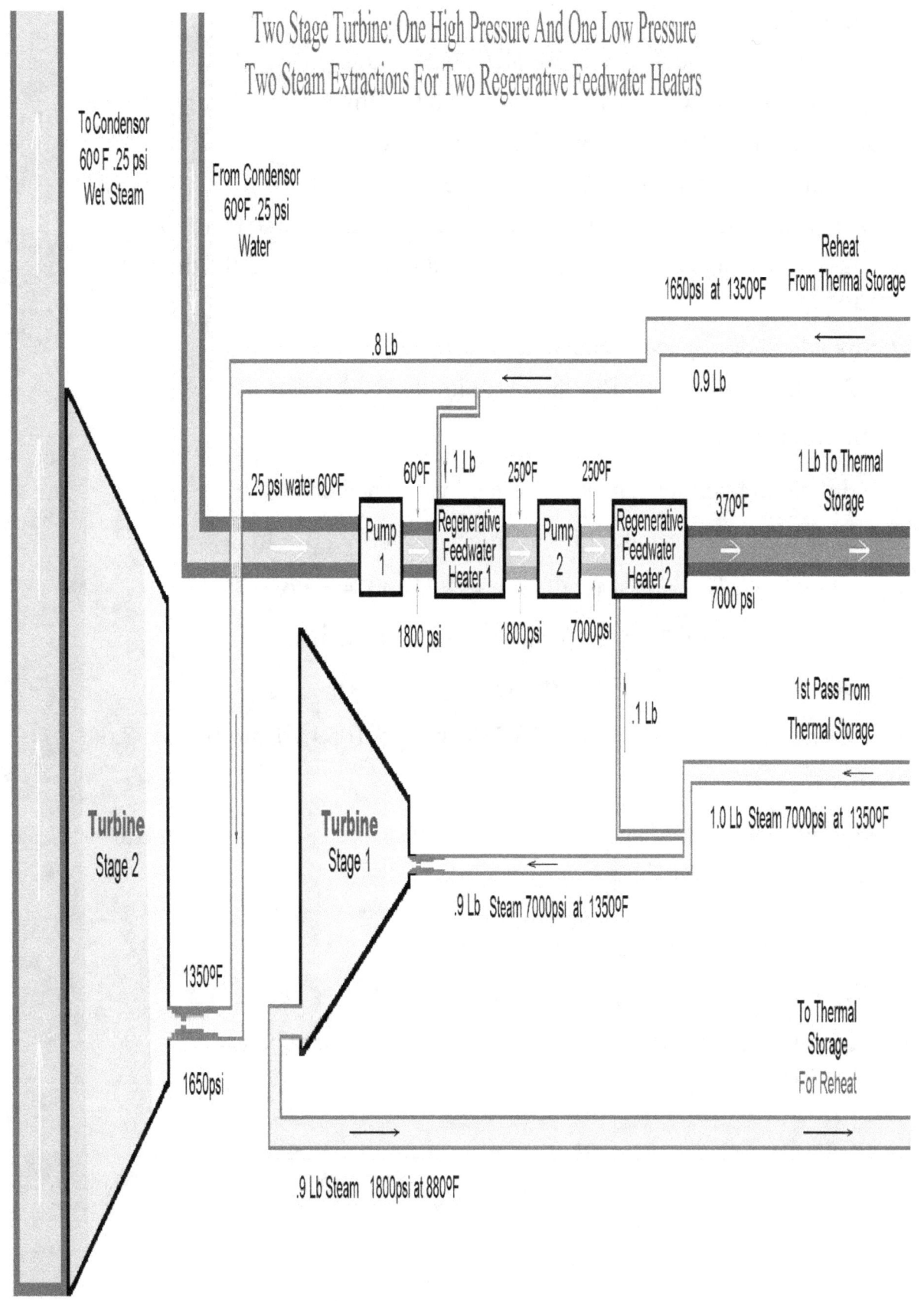

(.1 pound extracted for regenerative feed water heater)

.9 pounds steam at 7000psi at 1350°F at 1601.3 Btu/lb, S = 1.5152 Btu/lb°R

.9 lb expanded to 1800psi at 880°F at 1403.8 Btu/lb

Ideal work = .9lb x (1601.3 Btu/lb - 1403.8 Btu/lb) = 177.75 Btu

At 90% machine efficiency (windage, gaps, and etc.) = 159.89 Btu

17.77 Btu internal reheat on .9 lb steam heating it to 905°F

Extract .9 lb steam from turbine & reheat in thermal storage to 1350°F at 1650 psi (10% pressure loss extraction and reheat) at 1692.8 Btu/lb S = 1.7149 Btu/lb °R

Energy added during extraction and reheat

(1692.8 Btu/lb - 1403.8 Btu/lb) x .9 lb = 224.1 Btu

Second Expansion :

(.1 pound extracted for regenerative feed water heater)

.8 lb expanded to .25psi at 60°F (average) at 1087.7 Btu/lb

S_{gas} = 2.094 $S_{exhaust}$ = 1.7149 (2.094 - 1.7149) = .379

S_{fg} = 2.039 .379/2.039 = 18.59% wet steam at exhaust

Saturated steam .25psi 1059.7 Btu/ lb

Energy in 18.59% wet = .8 lb x 18.59% wet x 1059.7 Btu/ lb = 157.6 Btu/lb

Energy in dry expanded steam = (1087.7 Btu/lb x .8 lb) - 157.6 Btu = 712.6 Btu

Energy At Start Of Second Expansion :

.8 lb x 1692.9 Btu/lb = 1354.32 Btu

Work second expansion = 1354.32 start - 712.6 Btu exhaust
= 641.72 Btu

Total Work = 159.98 Btu 1st expansion + 641.72 Btu 2ed expansion = 801.7 Btu

First Regenerative Feed Water Heater

.8 lb from condenser at 60°F pumped to 1800 psi at 60°F at 33.1 Btu/lb

Energy available to condensate from 2ed extraction cooling from 1350°F at 1692.9 Btu/lb to
250°F at 222.2 Btu/ lb

.1 lb x (1692.9 - 222.2) = 147 Btu regenerative heat

147 Btu/ .8 lb = 183Btu/lb + 33.1 Btu initial = 216.93 Btu/lb at 250°F for mixed .1 lb 2ed extraction and .8 lb condensate = .9 lb

Second Regenerative Feed Water Heater

.9 lb at 250 °F pumped to 7000 psi at 233.3 btu/lb

.1 lb 1st extraction at 1601.3 Btu/lb cooled to 370 °F at 353.5Btu/lb

.1 x (1601.3 - 358.5) = 124.78 btu/lb regenerative heat

233.3 btu/lb +124.78 btu/lb = 358 Btu/lb at 370 °F

Energy Added From Thermal Storage

First Pass

7000 psi water at 370 °F at 358 Btu/lb pumped into thermal storage, heated to 1350 °F
at 1601.3 Btu/lb

(1601.3 Btu/lb - 358 Btu/lb) = 1247.8 Btu added first pass

Energy added turbine extraction and reheat 224.9 Btu (see above)

Total Energy Added From Thermal Storage

1247.8 Btu first pass + 224.9 Btu extraction and reheat second pass = 1472.7 Btu

Turbine Efficiency

Total Work = 801.7   Total energy added from thermal storage = 1472.7 Btu

Efficiency (before pump work) = total work/total energy added = 801.7 Btu/1472.7 Btu = 54.43%

Pump Work

1st pump (.25 psi to 1800psi)

Water at 60°F at .25psi at 28.1 Btu/lb pumped to water at 60°F at 1800psi at 33.1 Btu/lb

.8 lb / (.8 efficient pump x .8 efficient motor) x (33.1 - 28.1) = 6.25 Btu

2^{ed} pump (1800psi to 7000psi)

Water at 250°F at 1800 psi at 216.93 Btu/lb pumped to water at 250°F at 7000psi at 233.3 Btu/lb

.9 lb / (.8 efficient pump x .8 efficient motor) x (233.3 - 216.93) = 15.469 Btu

Total Pump Work = 6.25 Btu + 15.469 Btu = 21.719 Btu

Turbine Cycle Efficiency With Pumps ;

$$\frac{\text{Work Extracted - Pump Work}}{\text{Heat Added Thermal Storage - Pump Work}} = \frac{801.7 - 21.719}{1472.7 - 21.719} = 53.78\%$$

Thermal - Electrical Efficiency

53.78 % turbine cycle x .97% turbine mechanical eff. x 93.2% Generator = 48.619%

Internal Solar Electrical Production

1660.41025 Kw$_{SD}$ = Total Direct Solar Energy Intercepted By 45 meter Unit

1275.82 $Kw_{NetFromThermalStorage}$ (on an solar hour average basis - see above)
After 31 days storage at 0°F
x 48.619% thermal electrical efficiency = 620.2909 Kw_E

Additional Parasitic Losses

Electromagnetic Sodium Pump

Recall:

$\qquad$ 1414.263 Kw_{TEA} thermal energy absorbed at ceramic surface
$\quad$ -96.00 Kw_{RadSB}
$\quad$ - 11.13 Kw_{ConvSB}
$\quad$ - 4.01 Kw_{CondSB}
$\quad\underline{- 6.88\ Kw_{Sodium\ Line}}$
$\qquad$ 1296.25 $Kw_{NetToThermalStorage}$

However the sodium mass flow thru the boiler has to be enough to make up for the losses in the sodium line from the solar boiler until it enters the thermal storage system, previous losses are prior to energy being absorbed by sodium.

1296.25 $Kw_{NetToThermalStorage}$ + 6.88 $Kw_{Sodium\ Line\ Losses}$ =

=1303.73 $Kw_{EnergyFor\ SodiumMasFlow}$

Sodium Heat Of Vaporization = 1718 Btu / Lb

$$\frac{3413\,Btu/KwHr}{1718\,Btu/Lb} =$$

= 1.9866 Lbs Sodium vaporized per Kw_T To Thermal Storage

1303.73 Kw_T x 1.9866 Lbs =

2589.99 Lbs Sodium Vaporized Per Hour = Mass flow

Round Up To 3000 Lbs / Hr = 50 Lbs / min = .8333 Lbs/sec

Sodium Liquid Density = 46 Lbs/ Ft^3

Volume /Hr = 3000 Lbs / Hr /46 Lbs/Ft^3 = 65.27 Ft^3 / Hr

= 1.087 Ft^3 / Min = .01812 Ft^3 /sec

Maximum head[+] at noon June 21 less than 47 meters.
Round up to 50 meters = 165 feet

[+] Distance from sodium condensate reservoir to top of boiler

Horsepower $_{Head}$ =

$$= \frac{Head \times Density \times Volume/Sec}{550 \text{ foot/lbs/sec} \times \text{efficiency of sodium pump}}$$

Horsepower $_{Head}$ = $\frac{165 Ft \times 46 Lbs/Ft^3 \times .01812 Ft^3/sec}{550 \text{ foot/lbs/sec} \times .5 eff}$ =

= .5001 Hp

Derivation of friction losses in sodium feed, sodium boiler, sodium vapor to storage and condenser not shown taken as 5 times head horsepower = 2.5005 Hp = 1.869 Kw$_{ESP}$

Unit works 12 hrs/average day while sun is up. Due to small size and mass of solar boiler warm up time is short so working temperature is reached in relatively short time. However maximum mass flow requires maximum direct solar exposure available only for an average of seven hours per day. For purpose of discussion maximum mass flow will be based upon a 12 hour per day sodium pumping cycle. This means that 1.869 Kw$_{ESP}$ will equal the loss for the sodium pump.

Solar Electrical Output = 620.2909 Kw$_E$ - 1.869 Kw$_{ESP}$ =

+618.421 Kw$_E$ After Sodium Pump

Solar Clock Drive

Un-optimized non trussed equatorial yoke, and reflector support using standard A-36 structural steel supporting aluminum reflector designed for 100 mph wind at 60% of yield strength weighs approximately 1,047,739 lbs. Said mass is set to rotate about equatorial axis parallel to earth's polar axis once per day. Mass is not lifted but reflector is cradled in yoke assembly at balance point. However to err on the higher energy use side, assume clock drive must do the work of lifting said mass to a height of 25 meters each day.

25 meters x 3.2808 ft/meter x 1,047,739 lbs =

=85,935,553 foot pounds per day.

There are 24hrs/day x 60min/hr x 60seconds/min =

=86,400 seconds/day

$$\frac{85{,}935{,}553 \text{ foot pounds per day}}{86{,}400 \text{ seconds/day}} = 995 \text{ foot pounds/sec}$$

550 foot pounds/sec = 1 hp

$$\frac{995 \text{ foot pounds/sec}}{550 \text{ foot pounds/sec}} = 1.81 \text{ hp}$$

746 watts/hp x 1.81hp = 1,351 watts

With friction and minute corrections to adjust solar tracking to within 1/9 degree accuracy take solar clock energy at 1550 watts.

At 80% efficient electric motor take tracking electric consumption = 1.94 $Kw_{SolarClockDrive}$

As clock drive as envisioned runs 24 hrs per day

The total clock drive consumption would be:

$$\frac{1.94 \text{ Kw}_{SolarClockDrive} \times 24\text{hr/day}}{7.07 \text{SolarProduction Hrs}/24\text{hrday}} = 6.59 \text{ Kw}_{ClockDriveEquivalent}$$

618.421Kw_E After Sodium Pump $- 6.59 \text{ Kw}_{\text{ClockDriveEquivalent}} =$

$= 611.831 \text{ Kw}_E$ After Clock Drive

Lighting

Unit must be lit 12 hours per average day

Light on North Support, Bottom Of Trans Yoke beam, Both Reflector-Yoke Gimbals And South Support, with five 80 watt lamps

12 hrs/ave day x 5 lamps x 80 watts = $4.8 \text{ Kwhr}_{E \text{ Lighting}}$

$$\frac{4.8 \text{ Kwhr}_{E \text{ Lighting}}}{7.07 \text{SolarProduction Hrs /24hrday}} = .679 \text{ Kw}_{\text{LightingEquivalent}}$$

611.831Kw_E After Clock Drive $- .679 \text{ Kw}_{\text{LightingEquivalent}} =$

$= 611.152 \text{ Kw}_{E \text{ Net After All Parasitic Losses}}$

Solar Electrical Efficiency PILOT PROJECT**

$$\frac{611.152 \text{ Kw}_{E \text{ Net After All Parasitic Losses}}}{1660.41025 \text{ KwSD}_{\text{Total Direct Solar Energy Intercepted}}} =$$

$= 36.8072 \%$ Solar Electrical Efficiency $_{\text{Net}}$

U.S. SOLAR POWER SUPPLY

Efficient Solar Electrical Power Production With *Long Term Energy Storage* REPLACES Conventional Power Plant: With **Least** Reflective Surface

U.S. Solar Power Supply City Sized System

With Six Months Long Term Thermal Energy Storage

Summer To Winter At Continuous Average Demand Output

A U.S. Solar Power Supply City Sized System gathers more energy in the summer and less in the winter due to the normal seasonal fluctuation in solar energy available. In order to provide continuous power within the normal daily demand fluctuations, the U.S. Solar Power Supply City Sized System stores the surplus energy gathered during the summer for use during the winter, or in air condition dominated service sectors, it stores the surplus energy gathered during the winter for use during the summer.

Imagine an array of four-hundred-and-sixty-eight U.S. Solar Power Supply collectors each with a dish diameter of one hundred meters, arranged in a hexagonally closest packing grid where the edge of each dish is sixty-one and a half meters from its nearest neighbor.

Output From A Single Collector

The hundred meter collector as pictured page 18 was designed with a surface tolerance of plus or minus 50 Microns in order to keep the reflected radio waves in phase at the focus. Tolerances for solar need not be so stringent or costly. A hundred meter unit intercepts 8199.55 Kw_s of direct solar energy on a clear day, or 4.93 times as much as the forty-five meter prototype unit. The large unit's efficiency at capturing the solar energy in the boiler as thermal energy is the same as the smaller unit. 6451.91 Kw_T is retained in the sodium vapor in the boiler. The four-hundred-sixty-eight units gather a total of 3,019,493.88 Kw_T

Network Energy Losses

The sodium vapor is collected from four-hundred-sixty-eight collectors via a network of well insulated pipes. These pipes loose 11,156.51 Kw_T , plus daily warm up and cool down losses of an additional 50% equivalent for a total of 16,734.76 Kw equal to .5542% leaving 3,002,759 Kw$_T$ for deposit into thermal storage.

Thermal Storage Size

The total annual winter season deficit when compared to the average daily output for the U.S. SOLAR POWER SUPPLY CITY SIZED SYSTEM is equivalent to approximately thirty average days' output. To be on the safe side, the LONG TERM THERMAL ENERGY STORAGE will be sized to hold forty days' average output: And to store that amount of energy from summer to winter, or from winter to summer (six months) when it will be required.

3,002,759 Kw$_T$ x 7.06 Sunny Hrs/Ave. Day x 40 Days =
= 847,979,175.5 Kwhr$_T$

847,979,175.5 Kwhr$_T$ / 8.43 Kw$_T$ /Molten Ft$_3$ NaCl =
= 100,590,659 Ft$_3$ NaCl

Dividing volume required into four tanks 290 Ft on an edge yields :

Heat Loss From Four Tanks At 0 °F Ambient Air (Four Layers Of Insulation)

4 Tanks x 24Hrs/Day x 183 Days x 6 Walls x 290 Ft x 290 Ft x

$$\frac{1.5 \text{ Shape Factor} \times (1461°F - 0°F)}{\frac{3Ft}{.1 \frac{Btu}{HrFt°F}} + \frac{3Ft}{.05 \frac{Btu}{HrFt°F}} + \frac{3Ft}{.015 \frac{Btu}{HrFt°F}} + \frac{25Ft}{.01 \frac{Btu}{HrFt°F}}} =$$

= 6,352,876,184 Btu = 2,244,208.7 Kwhr$_T$ =

847,979,175.5 Kwhr$_T$ To LONG TERM THERMAL ENERGY STORAGE
- 244,208.7 Kwhr$_T$ Lost In 6 Months At °F

= 845,734,966.8 Kwhr$_T$ Left in LONG TERM THERMAL ENERGY STORAGE

Which is only .26465% Loss After Six Months

3,002,759 Kw$_T$ per average solar hour to LONG TERM THERMAL ENERGY STORAGE from 468 100 Meter Units - .26465% Loss =

= 2,994,8,12.2 Kw$_T$ to turbines

Thermal-Electrical Efficiency

The turbine cycle efficiency of the U.S. SOLAR POWER SUPPLY CITY SIZED SYSTEM will be the same as for the 45 meter prototype. The friction factor in the much larger CITY SIZED turbines will be only 1%. The CITY SIZED electric generators will be 98% efficient.

53.78% Turbine Cycle x 99% Turbine (1% Friction) x 98%.generator
= 52.177% Thermal Electrical Efficiency

Overall Solar-Electrical Efficiency

2,994,8,12.2 Kw to turbines x 52.1777% Thermal-Electrical Efficiency = 1,562,624.1 Kwe

$$\frac{1,562,624.1 \text{ Kw Electrical Energy Produced}}{3,837,389.4 \text{ Kw Solar Energy Intercepted}} = 40.721\%$$

Yearly Electrical Output From
U.S. SOLAR POWER SUPPLY CITY SIZED SYSTEM

8199.55 Kw$_s$ Solar Intercepted Per Unit, x 468 Units x 7.07 Sunny Hrs/Ave. Day Including Clouds x 365.25 Days Per Year x 40.72% Solar-Electrical Efficiency =

= 4,031,236,672 KwHrs Electrical Per Year

The Area Of A U.S. SOLAR POWER SUPPLY CITY SIZED SYSTEM

Using 468 of the hundred meter units arranged in twelve hexagonal rings about the LONG TERM THERMAL ENERGY STORAGE, with each unit's edge 61.5 meters from its nearest neighbor's edge occupies 4.47 Square Miles

Effelsberg 100 Meter In Diameter Radio Telescope is located at the Max Plank Institute in Germany is one of the largest fully steerable radio telescopes on earth. Since operations started in 1972, the technology has been continually improved (i.e. new surface for the antenna-dish, better reception of high-quality data, extremely low noise electronics) making it one of the most advanced modern telescopes worldwide. An almost perfectly shaped smooth surfaced paraboloidal radio telescope 100 meters in diameter was built at the Max Plank Institute whose surface is accurate to one quarter wavelength, and corrects for difference in phase for waves reflected from the center and edges of the reflector due to further distance while also correcting for different phase resulting from waves arriving at different angles.

U.S. SOLAR POWER SUPPLY
Efficient Solar Electrical Power Production With– *Least* Fluid Mass Flow / $Kwhr_E$

U.S. Solar Power Supply National Sized System

With Six Months Long Term Thermal Energy Storage
Six Months At Continuous Average Demand Output

Sized To Meet 100% Of All United States Present Electrical Demand: Equal To All The Power Now Produced (2015) From Coal, Nuclear, Natural Gas, Oil, Hydroelectric, Wind, Including All Private Power Producers

The Total Electric Consumption ForThe United States in 2017*** was 3,911 Billion Kwhr$_E$

One U.S. Solar Power Supply City Sized System occupies 4.47 Square Miles and produces 4,029,383,155 KwHrs Electrical Per Year.

To supply 100% of today's United State's electrical demand would require :

$$\frac{3911 \text{ Billion Kwhr}_E \text{ Total US Electrical Demand ***}}{4,031,236,672 \text{ Kwhr}_E \text{ U.S. SOLAR POWER SUPPLY* CITY SIZED SYSTEM ** Output}}$$

= 970.173 U.S. Solar Power Supply City Sized System
to Meet Total Current US Electrical Demand

At 4.47 Square Miles Per U.S. Solar Power Supply City Sized System the total needed for a true U.S. Solar Power Supply would be :

= 970.173 x 4.47 square miles =

= 4,335.9 square miles to power the entire United States

***CIA World Fact Book 2016

4,335.9 square miles is a square 65.84 miles on a side.

U.S. Solar Power Supply National Sized System
Area Needed To Equal Present Demand [DOE 2017]

Total Area Apportioned Among The Several States Of Arizona, California, Colorado, Nevada, New Mexico, Texas, and Utah. Each Area Is Slightly Less Than 25 Miles by 25 Miles.

The United States has an area of 3,615,122 square miles.

$$\frac{4{,}339.9 \text{ square miles needed}}{3{,}615{,}122 \text{ square miles}} = .0012005$$

= just over 1/10 of 1 % of United States Area 3,615,122 square miles

U.S. SOLAR POWER SUPPLY NATIONAL SIZED SYSTEM
With Six Months LONG TERM THERMAL STORAGE ** Duration:
At Continuous Average Demand Output

Including All Expected Growth In Demand Until 2050

If United States electrical growth were held to 2% per year including the introduction of increased use of electric cars, the power required in the year 2050 would be :
(starting from DOE 2017 demand)

$(1.02)^{50 \text{Years}} = 2.6915$ times more power

Area Of 2050 U.S. Solar Power Supply National Sized System

4,111.67 square miles present demand area x 2.6915 =

=11,066.56 square miles

Requiring that much more land equal to a square about 105.2 miles on a side 2.6915 x .0011374 Of United States Land Area (2017 Area Of Land Needed) =.0030613
 = Just At 3/10ths of 1 %

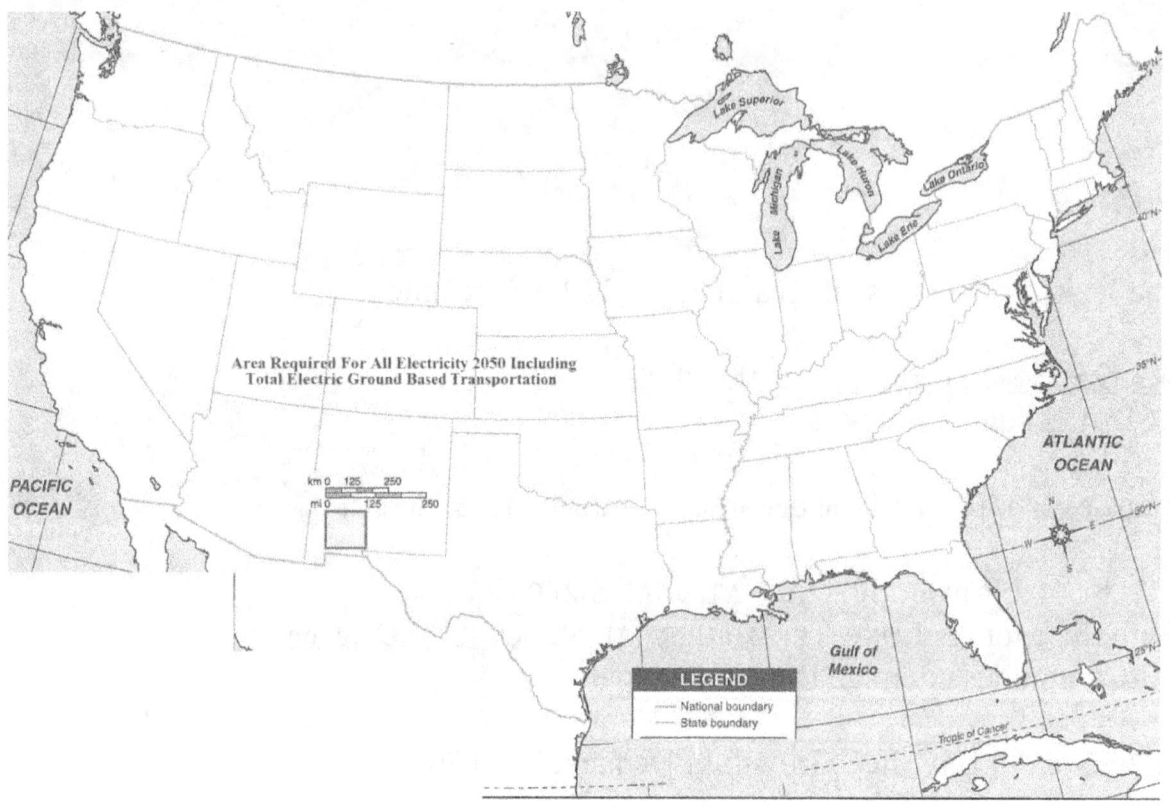

The Case For Long Term Thermal Energy Storage

Up until now, the capital cost of a solar power system - *without* Long Term Thermal Energy Storage - could only be written off against the fuel savings of the conventional power plant it supplemented. To provide power to a given service area, in essence, that service area had to bear the capital cost of both the conventional power plant and the more expensive solar power plant as well.

However with its LONG TERM THERMAL ENERGY STORAGE the U.S. SOLAR POWER SUPPLY replaces a conventional power plant, rather than merely reducing the conventional power plant's fuel cost. As a consequence the capital cost of U.S. SOLAR POWER SUPPLY can be written off against the savings in the capital cost of the conventional power plant it replaces.

The U.S. SOLAR POWER SUPPLY can stand alone.

Scenario A

For extremely hot locations like Tucson Electric Power's service area, peak summer air conditioning drives their average annual load factor down to around 40% compared to the national load factor of 62.5% (D.O.E. 1999). With U.S. SOLAR POWER SUPPLY, the spring, fall, and winter surplus can be stored for use during the summer peak season reducing the size of the required solar collector.

Scenario B

There is contemplated a proposed mandate that by 2010, 10% of power production in certain areas be provided by alternative energy sources. According to Colonel Boardman, Department Of Defense Army Installation Management, military bases are already under that mandate. It is also hoped to increase that dependence to 15% by 2015. Many envision that reflecting trough collectors, central receivers, along with some photo voltaic systems would provide this energy. (Was not met)

There have been periods of cloudiness over widespread areas of the Southwest lasting four to seven days, although it happens infrequently. Less unusual are periods when series of storms roll in from the Pacific blanketing the region with clouds for one to three days followed by one or two days of sun followed by another period of one to three days cloudiness.

A utility that serves a given service area must have a dependable, conventional, installed capacity equal to the maximum anticipated demand plus safety factor. Without Long Term Thermal Energy Storage those utilities that have become dependent upon solar energy providers for any actual percentage of their demand with only short term storage will eventually face a shortfall with serious consequences.

Scenario C

Places like San Diego, Los Angeles, Las Vegas, Reno, Phoenix, Tucson, Albuquerque, El Paso, and Santa Fe, along with the smaller cities near them have power demand profiles that more or less match the seasonal fluctuation characteristic of non - Long Term Thermal Energy Storage solar systems operating from the Southwest. However, there are many Utility Service Areas located just to the northwest, north, northeast and east of the primary solar collection area that have demand cycles that are not in sync with the annual summer-winter solar cycle of the Southwest.

Without Long Term Thermal Energy Storage these areas cannot depend upon solar power for more than a few percent of their total demand limiting solar power to a minor role in The Southwest.

Scenario D

Utilities prefer to have an installed capacity equal to 110% to 115% of their record maximum peak demand. For a place like Denver, that peak demand might occur in the early part of January during the second or third day of a blizzard when its -15°F. Under such conditions all the lights are on, both because the days are short, and also because it's dark even during the day. All the heat pumps in the service area are on emergency backup electric resistance heat and because the houses' heating systems were optimized for 10°F - to save money - some homes have one or two extra space heaters going. It's one thing to have a brownout in Southern California during the summer, a lot of people are going to sweat. But for most utilities in the United States, if the power fails on that cold January day when they truly need it the most, its life and death.

In the case of a primarily coal fired utility system, like Excel Energy Of Colorado for example, the utility executives like to have on hand, in the yard, a several day coal reserve for the grinders/blowers, a hundred car coal train pulling into the yard, another hundred car coal train on its way to arrive in ten to twelve hours, and another one being loaded at one of the mines they rely on.

If a broad spectrum of utilities are going to use solar power to any large extant, they want a system that has that equivalent of the pile of coal in the yard with another 100 car coal train on the way. The U.S. SOLAR POWER SUPPLY with LONG TERM THERMAL ENERGY STORAGE is that equivalent.

Let's face it, forcing utilities to pay 8, 10, 12 cents a $Kwhr_E$ for intermittent power that they can't count on when they really need it, just isn't right. If intermittent solar power dependence should grow to more than 4 or 5%, it's a calamity waiting to happen.

Mr. Tex Wilkins, DOE Solar Thermal for twenty-five years, admitted to me that during his watch they have spent in excess of 4.6 billion dollars on solar thermal power. Yet they have never built one solar system that will store one season's surplus for use during another season's deficit. Intermittent solar energy is a fair weather friend whose eventual betrayal will be catastrophic. And those who provide real power to sizable service sectors know it.

Many have toiled for many years, given their whole careers to solar energy - in order to really make a difference - both in the air we breathe and the geopolitical conditions we live in. I don't mean to denigrate or diminish their work. But without LONG TERM THERMAL ENERGY STORAGE solar power as a percentage of total power production will always be in the single digits, and these people's dreams and aspirations will go largely unrealized.

U.S. SOLAR POWER SUPPLY

Efficient Solar Electrical Power Production With *Long Term Storage* REPLACES Conventional Power Plant – And All Its Infrastructure

The Case For Maximized Solar/Electrical

Conversion Efficiency

Just for a moment or two, let us entertain an entirely different point of view, an analogy to solar energy conversion, namely rocket science. In rocket science, a certain minimum set of parameters have to be met in order that a payload achieve orbit. If you don't met these criteria, you have an Alan Shepard falling into the water a couple hundred miles of the coast of Florida, and you let the media shout hurrah long and loud enough to claim a victory to the people who paid for it, but everybody in the know, knows he's no Gragarin.

Now let me ask you a question. If you were going to build a rocket, would you use gasoline as the fuel and compressed air as the oxidant? Would you use gasoline and liquid air? In order to maximize the thrust per pound of fuel and oxidant - to get the highest specific impulse - you would chose liquid oxygen and liquid hydrogen. But liquid oxygen and liquid hydrogen are more expensive. For 20 years plus, the Federal Government figuratively speaking, has spent 3.5 billion funding gasoline and liquid air. One of my references, who will remain unidentified says, "No, they chose wood and air."

Let us consider this rocket analogy from another perspective for a moment. Let us say we can chose from between two rocket engines. Rocket engine "A" is 50% efficient at converting the combustion energy in its fuel and oxidant into thrust, and rocket engine "B" is 25% efficient at converting the combustion energy in its fuel and oxidant into thrust. Now let us further suppose that the scientists who are going to decide look at the more costly materials requirements of the more efficient rocket engine "A" that operates at higher temperatures and pressures and make a decision to chose rocket engine "B" to save money.

OK, they have saved money on the rocket engine, but what are the consequences. First of all, the less efficient rocket engine "B" must burn fully twice as much fuel and oxidant to develop the same thrust. However, it must necessarily develop twice as much thrust to get off the ground because the weight of the fuel and oxidant are twice as heavy and the weight of the propellants' containers must necessarily weigh twice as much because strength (section modulus) and stiffness (deflection, i.e. young's modulus) fall off as $1/x^2$ and $1/x^3$ respectively.

So, not only must the less efficient rocket "B" burn twice the fuel to lift the same weight, it must also burn twice as much again OR FOUR TIMES MORE ALTOGETHER to lift twice the weight and it must burn at a rate of at least twice as fast as the more efficient rocket engine "A". Since it must burn at twice the rate, the size and actual weight of the engine itself must be more to handle the doubling of the throughput volume. Since the engine is bigger and subject to the same strength and deflection laws as the tanks it also weighs more, and it has to lift its own greater weight also - which requires more thrust, more fuel, and incrementally larger propellant tanks once more. In all likelihood the propellant tanks will be wider offering more wind resistance increasing the thrust, fuel, oxidant, tank size, and engine size requirements once again.

In all certainty, if one chooses the rocket engine that is half as efficient results in a rocket that is more than twice as heavy.

The same relationships hold true for solar energy conversion. Lowering the efficiency of any of the eighteen steps of energy capture and conversion effects not just its own coefficient in the process, but also negatively effects many of the other seventeen coefficients of conversion as well.

For example: Lowering the temperature of storage lowers the turbine efficiency which means that more energy has to be stored for the same net output. If more energy must be stored, more must be collected so the pipe carrying the energy must be wider and have more surface area to lose energy. Also, since more energy must be stored, more must be collected and the size of the collector must necessarily be larger. With a larger collector, the distance from the collector to the thermal storage tank must be a greater distance so the collection pipe must not only be wider but longer and lose even more energy. Collecting more energy means more fluid has to be pumped with greater pumping losses. Each one of these losses has to be made up for by gathering more energy and having a larger collector and longer wider collection pipe etc.

Lowering the temperature of storage lowers the turbine efficiency which means that more energy has to be stored for the same net output which requires a larger tank. The larger tank has a larger surface area, takes more material and has more surface area to lose heat from. All this requires a larger collector, and a longer fatter collection pipe etc.

The same thing happens if you don't track the sun dead on full normal, or use high concentration. The results snowball throughout the system resulting in lowered efficiencies for the coefficients all the way down the line. That's why a trough system takes between three and four times as much reflective surface area to generate the same total kwhrs during a given solar day, the central receiver requires between two and three times as much, the sterling dish up to about 50% more, and none of them has ever been coupled to a LONG TERM THERMAL ENERGY STORAGE
 that will free the unit from a conventional power plant backup, and as a group will always be relegated to the status of supplemental or intermittent power and never reduce the nameplate installed capacity of conventional power plants utilities can depend on.

The U.S. SOLAR POWER SUPPLY utilizing : Full normal solar tracking to get the maximum solar energy available, high concentration to minimize boiler and boiler support shadow losses - and reduce convection, conduction, and radiation losses, and thus achieve vaporization of sodium to transport the maximum amount of energy per surface area of boiler and reduce the mass flow and pumping losses to a minimum, allow said energy to melt common salt, NaCl, in storage to achieve the 8.2 $Kw_{Thermal}$ per cubic foot at 1480 °F which allows for energy available for the world's most efficient supercritical pressure steam turbines at over 1350 °F to reduce the size of the LONG TERM THERMAL ENERGY STORAGE to a minimum, to reduce collector size to a minimum, to reduce collection pipe length and energy losses to a minimum. Only the normal tracking, highly concentrating U.S. SOLAR POWER SUPPLY can achieve the status of base load power that can replace - not just supplement - conventional power plants.

U.S. SOLAR POWER SUPPLY vs. Photovoltaic &

Photovoltaic Distributed Generation

Current state of the art crystalline silicon solar cells run at 12 to 20% efficiency in converting solar energy to electricity. With reflectors mounted to concentrate the solar energy onto the surface of the cells they have reached 20% efficiency.

Comparing the very most efficient photovoltaic systems with the U.S. SOLAR POWER SUPPLY CITY SIZED SYSTEM efficiency at 40.7% we see that these photovoltaic systems would require slightly more than 2 times more solar collector surface area and land.

The above comparison assumes that the photovoltaic system would track the sun "normally" as the. U.S. SOLAR POWER SUPPLY
does. If the photovoltaic surface rested at some seasonally adjustable optimum fixed angle it would only receive 87% as much as the tracking system which would mean the photovoltaic system would require 2.29 times as much collector surface area and land as the U.S. SOLAR POWER SUPPLY

If the tracking concentrating photovoltaic system were to store the electrical output for cloudy days or night time periods the energy would be subject to a minimum 20% storage loss in the batteries and another 10% loss to get the required AC from the batteries' DC. Even the highest efficiency photovoltaic system with storage would require 2.77 times more collector surface and land area to generate the same ac power as the U.S. SOLAR POWER SUPPLY

Using a non-tracking, non-concentrating photovoltaic system with storage would require 4.33 times more collector surface and land area than the U.S. SOLAR POWER SUPPLY
.

If transient DC power is OK and these solar cells were in L. A. or some other less than optimal solar collection site area, then the non tracking PV (PC) systems will require 2.29/.8 local solar availability equal to 2.86 times the surface area. If they are not mounted at the ideal non tracking angle, say on a pleasing vertical or horizontal surface to meld into the buildings in some non-affronting posture, then you are looking at way over three times the surface area compared to the U.S. SOLAR POWER SUPPLY

. All this doubling and tripling of solar collecting area not to mention the exorbitant cost per actual $Kwhr_E$ produced just to save the 2% to 5% transmission line loss . . . it is not reasonable.

After the extra expense of buying all these none-too-cheep solar cells: On cloudy days the full weight of total demand for electrical power will be resting on the shoulders of the solar dependent and thus under capacity conventional power plants unless these solar cells are installed on any but the smallest - thus insignificant - scale. Eventually the customers in the service sector will rue the cloudy day.

U.S. SOLAR POWER SUPPLY vs. Cylindrical Parabolic
Trough Collectors And Central Receivers

Cylindrical Parabolic Trough Collectors

Back in the 1950's near perfectly shaped paraboloidal concentrators achieved solar concentrations of more than twenty-thousand to one producing temperatures at the focal point near that of the sun's surface.

In 1971 an almost perfectly shaped smooth surfaced paraboloidal radio telescope 100 meters in diameter was built at the Max Plank Institute whose surface is accurate to one quarter wavelength, and corrects for difference in phase for waves reflected from the center and edges of the reflector due to further distance while also correcting for different phase resulting from waves arriving at different angles. Others even larger sharing the same characteristics have been built.

While these 100 meters in diameter radio telescopes are also totally steerable horizon to horizon and pole to pole, the U.S. SOLAR POWER SUPPLY reflector is not a perfectly shaped paraboloidal, but produces a solar concentration of one thousand to one and merely follows the sun. Its solar concentration of one thousand to one allows it to operate at the very high temperature necessary for high thermodynamic efficiency at the turbine while still having enough thermal gradient or surplus temperature to melt salt, NaCl. at 1486°F.

On the simplest level, solar-thermal-electric systems utilizing either central receivers or cylindrical trough concentrators, due to their geometry, cannot achieve the high concentration ratios necessary to charge a low cost, high temperature, high efficiency system. This rules out both medium term and extended duration storage required for standalone base line service. These systems can only have expensive short term duration storage that merely reduces coal or other fuel use at a conventional power plant. The cost of the non-storing solar plant can only be written off against conventional fuel savings, not the capitol cost of the conventional power plant. These trough and central receiver systems can never stand on their own nor reduce nameplate capacity of the conventional power plant infrastructure.

Conventional power plants to meet peak demand are still required for extended cloudy periods or for seasons when demand is higher than daily output from solar. So the utilities and the end users still have to pay for both the conventional power plant, and the solar power plant that only provides intermittent power when the sun shines or for a short time thereafter. The customer still has suffer the associated pollution and support the conventional power plant's extended infrastructure such as mines, trains, pipelines, oil tankers, or sent fuel facilities.

The conventional trough concentrator cannot track the sun normally (totally perpendicular to the sun) to keep the area of interception perpendicular to the sun. So the trough concentrator does not receive as much energy the U.S. SOLAR POWER SUPPLY totally normal tracking reflector. Not getting the maximum amount of sun requires that the whole system be made bigger to catch more sun to make up the difference.

The cylindrical parabolic trough systems have solar concentration ratios of only forty to eighty. Higher concentration troughs (up near near eighty solar concentration) use deeper troughs that require several times more reflective surface than the normal interception area perpendicular to the sun.. The reflective surface of the U.S. SOLAR POWER SUPPLY is only 25% more than the normal interception area due to the slight curvature of the paraboloidial.

Because cylindrical trough concentrators' concentration ratio is so low, and the focal point is actually a line, not a point, the trough concentrator has a line boiler that runs down the axis of the trough, not a point focus like U.S. SOLAR POWER SUPPLY.

The trough concentrator's line boiler casts a ten times larger shadow letting less light get to the reflector and less light to the boiler reducing energy input even more. This reduces overall efficiency requiring that the whole system be made bigger to catch more sun to make up the difference.

As the boiler is inherently larger with an aperture area at a minimum ten times larger than the U.S. SOLAR POWER SUPPLY
approximately ten times as much energy is lost to radiation, convection and conduction losses at a given operating temperature. The trough collectors cannot achieve the very high temperatures needed for high thermodynamic efficiency after storage so cylindrical trough concentrator systems can only be used to reduce fuel use of a conventional power plant, never truly stand on its own.

Because trough concentrator systems operate at lower temperature with lower thermodynamic efficiency at the turbine, more energy must be transported to the power plant for a given amount of net electrical output. Collecting more energy requires more collectors and a larger field of collectors that produce more heat transfer fluid that must be transported from collectors to power plant. Not only is more fluid required from more collectors, but it has to be transferred a greater average distance. Because the cylindrical trough concentrating system operates at lower temperatures, and can't achieve the high temperature to use sodium vapor as its heat transfer fluid, the system must use another fluid carrying less energy per mass and volume. So the amount of heat transfer fluid must be increased again to make up for the difference and still carry enough energy to produce equal output.

This requires more energy for pumping. Because more energy is used for pumping, the system is even less efficient, and even more collectors are needed for the same output requiring the heat transfer fluid move a greater distance to the power plant requiring more energy. Because the pipes have a longer distance to go, they have more surface area and lose more energy. Because they are longer they have more mass and suffer more heat up losses in the morning, and more cool down losses in the evening. The whole system is less efficient requiring even more collectors to make up the losses requiring more mass flow, more pumping losses etc.

While the cylindrical trough concentrating system collector cannot replace a conventional power plant, they will take up three to four times as much land area as the U.S. SOLAR POWER SUPPLY power plant producing the same net electrical output. The trough system will also use between four and a half and five times as much reflective material on its array of collectors as the U.S. SOLAR POWER SUPPLY power plant producing the same net electrical output regardless of storage or lack thereof.

Central Receiver Collectors

Utilizing an array of solar tracking mirrors reflecting intercepted sunlight into a fixed solar boiler supported by a tower, central receivers can achieve high concentration ratios and high temperatures. A critical drawback however is that the solar tracking mirrors must track the half the angle between the sun and the boiler. Half the available solar energy comes between 9:30 A.M. and 2:30 P.M. The angle between the sun and the tower boiler in order to track at 9:00 A.M. or 3:00 P.M. is a minimum of 45 Degrees. Because the mirror is tracking the angle between sun and boiler it does not face directly at the boiler or the sun. So less solar energy is intercepted than a true normal tracking surface.

Because the boiler is fixed and sun moves, individual reflecting mirrors require space between them so they won't physically impinge upon each other. They need to be moved even further apart so that they won't shadow each other. The radius of the mirror field for the central receiver has twice the radius of the U.S. SOLAR POWER SUPPLY paraboloidal reflector for equal solar energy interception at noon in the winter. Before or after noon would require even more space for equal solar interception.

Because the individual reflecting mirrors are further apart, the image of the sun they reflect is larger which means that the boiler has to have a larger aperture area to let in the combined images of the sun focused by the solar array. The boiler has to be big enough to let in the reflection from the reflector sited at the greatest distance at the periphery of the array. Assuming equal focusing power for each independently moving reflecting element and the individual triangular reflecting element of the composite paraboloidal

U.S. SOLAR POWER SUPPLY the image area of the central receiver will have twice the radius and four times the area.

Unless the boiler is also twice as high, which would also increase the area of the focused beam another four times, the angle between the independently moving reflecting element and the horizontal plane of boiler aperture that must admit images from the entire field of mirrors, spreads the image out to a larger area dependent upon the angle of each beam from the disparate reflectors.

All these factors, plus several others too detailed to enumerate here act to increase the central receiver's boiler to a *minimum* of 6.28 times U.S. SOLAR POWER SUPPLY boiler size to capture the same energy at noon. Before 9:00 A.M. and after 3:00 P.M. these geometrical conditions increase the solar image size even more while mutual shadowing between the individual reflecting mirrors reduces reflected solar energy to the boiler.

On the average day, just slightly less than half the sun's total energy is Before 9:15 A.M. and after 3:15 P.M. As the angle increases between sun and boiler the reflecting mirrors present less area to the sun and reflect less also, even if they were not shadowed by adjacent mirrors. The central receiver efficiency drops for a large amount of available solar energy.

With a boiler six times larger for the same energy at noon and a larger shadow on the reflector elements, more energy is lost. Adding more reflective elements to the periphery of the array cannot add significant energy to the boiler without expanding the boilers aperture to admit the larger solar images reflected from the edge of the array. This would increase all the heat losses and cast more shadow losses. All the inefficiencies discussed in the section on trough collectors are to a lesser degree true of the central receiver. Experience has shown that the central receiver system cannot achieve high efficiency capture of solar energy during significant portions of the total solar day. Nor can it achieve high enough temperatures for the most efficient thermodynamic efficiencies at the turbine. Lastly, it cannot attain both high enough temperature to store energy long term and convert that energy at high efficiency.

Expressed Support

From: Wilkins, Frank [mailto:Frank.Wilkins@ee.doe.gov]
Sent: Thursday, March 05, 2009 2:29 PM
To: randyrosssolar@comcast.net
Cc: Ring, Bradley
Subject: U.S. Solar Power Supply
(Excerpted, Emphasis added)

Randy:

I looked through the project summary you left. A few thoughts about it:

* had you turned it into a proposal, it would have qualified for both of the concentrating solar power (CSP) solicitations we released in the last two years.

The first solicitation asked for innovative CSP concepts with an emphasis on storage. It resulted in 12 contracts. We didn't get as many good storage proposals as I hoped so we had a second solicitation focused on storage concepts. It resulted in 14 storage contracts and one heat transfer fluid contract.

* depending on how much funding we get in the FY2010 appropriation, there *may* be another CSP solicitation.

The development of low-cost CSP systems with storage is the goal of our program. There is no higher priority.

* your summary emphasizes efficiency and capability. I hope these comments help,

Tex

From Idaho National Lab: Formally Argonne West National Lab
- Site Of Breeder Reactor -

Sodium Used 30 Years
Re: Efficient Solar Electrical Power Production With Long Term Storage
Replaces Conventional Power Plant
Allows True Green Automobiles

Sent By:	"Sergiy V Sazhin" <Sergiy.Sazhin@inl.gov>	On:	May 05/14/09 7:00 PM
To:	RandyRossSolar@comcast.net		
Cc:	"Steven E Aumeier" <Steven.Aumeier@inl.gov>; "Timothy C Murphy" <Timothy.Murphy@inl.gov>; "Patrick W Kern" <Patrick.Kern@inl.gov>; "Steven D Herrmann" <Steven.Herrmann@inl.gov>; "Collin J Knight" <Collin.Knight@inl.gov>; "J A Michelbacher" <J.Michelbacher@inl.gov>; "Richard D Boardman" <Richard.Boardman@inl.gov>		

Dear Randy,

Now I am ready to answer your question you asked during our phone conversation regarding INL expertise in sodium metal in liquid and vapor states.

My managers recommended several experts to help you with your project. Please find their names with contact information below:

-Pat Kern, (208)- 533-7512, Patrick.Kern@inl.gov
-Steve Herrmann, (208)- 533-7859, Steven.Herrmann@inl.gov
-Collin Knight, (208)- 533-7707, Collin.Knight@inl.gov
-Bert Michelbacher, (208)- 533-7177, J.Michelbacher@inl.gov
-Charlie Griffin (Bert Michelbacher can provide contact information)

Have a success with your project.

Best regards,

Sergiy

Re: Efficient Solar Electrical Power Production With Long Term Storage Replaces Conventional Power Plant - Allows True Green Automobiles

Sent By: "Patrick W Kern" <Patrick.Kern@inl.gov> On:May 05/19/09 5:33 PM
To: RandyRossSolar@comcast.net
Cc "Sergiy V Sachin" Sergiy.Sazhin@inl.gov; "Van R Sandifer" Van.Sandifer@inl.gov

Mr. Ross,

First, my apologies for not sending this note earlier. Following our conversation
I fell Ill over the weekend and was unable to complete this correspondence.

Your proposal is intriguing in its departure from current approaches using the sun to provide renewable energy for our nation. There are of course some technical issues to address with the use of sodium, but I do not see any insurmountable obstacles from this high-altitude review of the papers you provided. As presented, the use of sodium as a heat-transfer medium to a sodium chloride storage bed provides
distinct advantages over currently available technologies.

I commend you on your novel approach and I will provide assistance as I am directed by my management to aid you in your task.

Patrick W. Kern
Space and Security Power Systems Facility
Materials and Fuels Complex
Battelle Energy Alliance

Mr. Jeff Goldberg, Dean of Engineering at University of Arizona
Put me in touch with one of his staff. Dr. Peiwen (Perry) Li.

Sent By: peiwen@email.arizona.edu   On: Jun 06/01/09 9:03 PM
To: To RandyRossSolar@comcast.net

Dear Randy:

Thank you very much for your following up email. As you know that I like the idea and is very interested in forming a team and work together in the near future.

I am traveling for an international conference. I will be back to Tucson next weekend. I will work on organizing a team and write a proposal as soon as possible.

Peiwen (Perry) Li
Assistant Professor
Department of Aerospace and Mechanical Engineering
The University of Arizona
Tucson, AZ 85721

Sent By: "Subbarao Surampudi"
subbarao.surampudi@jpl.nasa.gov On: Jul 07/15/09 4:33 PM

To: RandyRossSolar @ comcast.net

Dear Randy:

 Thanks for leaving a message with your e-mail. I had problems because I was putting a period between your first and last name.

 As I spoke to you earlier that my trip to DC was cancelled as my boss is having some issues.

I like the idea of solar thermal power plants and thermal storage. JPl played a significant role in this area twenty years ago. Presently we are working on novel laten heat thermal storage methods under a DOE sponsored program

I would to meet you soon and learn about your new concepts and explore the possibility to work together on a new proposal to DOE or other agencies.

 I may come to DC within two weeks. I will let you my travel plans once they are finalized.

 Looking forward to meeting you soon.

 Thank you very much

Rao surampudi

Mr. William Stein
CEM, CEP, Energy Coordinator
Fort Huachuca, Arizona

"This proposal and project has excellent merit and potential for giving this country an alternative source of electricity to fossil fuels or nuclear power. With the low cost of storage, it solves the inherent problem with solar energy to electrical plants (intermittence). Having worked in the energy management field I have (used) a lot of different ideas, systems and (seen) a lot proposals. Of all of them, this one has the best potential for becoming commercially viable in a very short time and extremely beneficial to this country. It would be an honor for Fort Huachuca to be the site of the first prototype system."

Dr. Fred Morse
Former Program Director, D.O.E. Solar
President Solar Energy Industries Association
Concentrating Solar Power Association

"High temperature, high efficiency, long term storage, that's the direction we should be going in. I think you have a great idea. Randy, I've helped you as much as I could my own son."

Professor Jeff Tester,
MIT, Cambridge, Massachusetts
Outside D.O.E. Solar Consultant

"We need a new program investigating longterm storage as pertains to the mix of solar power systems and its overall effect of the general grid stability of power production."

Benjamin P. Riley
Assistant Deputy Under Secretary Of Defense
Advanced Systems & Concepts

"This is intriguing. We need power systems to support the troops that don't have to be constantly resupplied with fuel. Can you break it down and load it on a Hercules? You need to talk to DARPA"

Dr. Anthony J. Tether
Director of DARPA
Defense Advanced Research Projects Agency

"1350 °F input - 60 °F exhaust, impressive. What's that in Rankine? (Answered by staff) That's a Carnot efficiency of over 60%. (to Staff)"Check out his calculations" (To me) "We'll get back to you."

Dr. Steven G. Wax
Deputy Director of DARPA
Defense Advanced Research Projects Agency

"After checking out your materials (calculations), we see no technical flaws, but its (Solar Power Supply*) a D.O.E. project, not a DARPA project"

Mr. Jim Rannels
Outgoing Program Manager D.O.E. Solar Thermal,
Promoted To D.O.E. Building Self Sufficiency

"If you had brought this to me two years ago when I had the money and the power, I would have signed off on it immediately. But now I do not have the money, and I do not have the power.

Mr. Steve Traver
Aeronautical And Mechanical Engineer
Energy Committee And Assistant To Senator
Pete Domenici, Chair Of The Energy And Natural Resources Committee

"This is a paradigm shift in solar energy, a real break-through. There is not a doubt in my mind that it will work. All the materials are off the shelf. This is the most mature solar energy technology I've seen in all my years."

Land Trust
U.S. SOLAR POWER SUPPLY

Whereas it is desirable that the United States should have an abundant unrestricted source of energy for economic reasons

Whereas it is desirable that the United States should be energy sufficient and depend upon domestic sources of energy and not foreign sources for both economic and strategic reasons

Whereas it is desirable that the United States should avail itself of an abundant unlimited source of energy that pollutes neither air or water nor admits any greenhouse gases

And whereas there now exists a solar-electrical conversion technology with long term energy storage - not requiring fossil fuel backup-that meets these aforementioned criteria at more than 40% efficiency such that an area of land equal to ½ of 1% of The United States territory when fitted with said technology will be sufficient to provide all expected demand for electrical and ground based transportation energy through the year 2050

Therefore The Executive Branch now orders that an area equal to ½ of 1% of United States territory located within Federal Lands in the Great American Southwest apportioned among the several states of Arizona, California, Colorado, Nevada, New Mexico, Texas, and Utah shall now be set aside in the U.S. SOLAR POWER SUPPLY **Land Trust** to allow development of a true U.S. SOLAR POWER SUPPLY

Submitted Respectfully

Randal Ross

U.S. SOLAR POWER SUPPLY

Efficient Solar Electrical Power Production With *LongEnergy Term Storage*

Distributed Area U.S. SOLAR POWER SUPPLY Land Trust To Meet Total 2050 United States Electrical Demand Including An **All Electrical Ground Based** -Transportation System

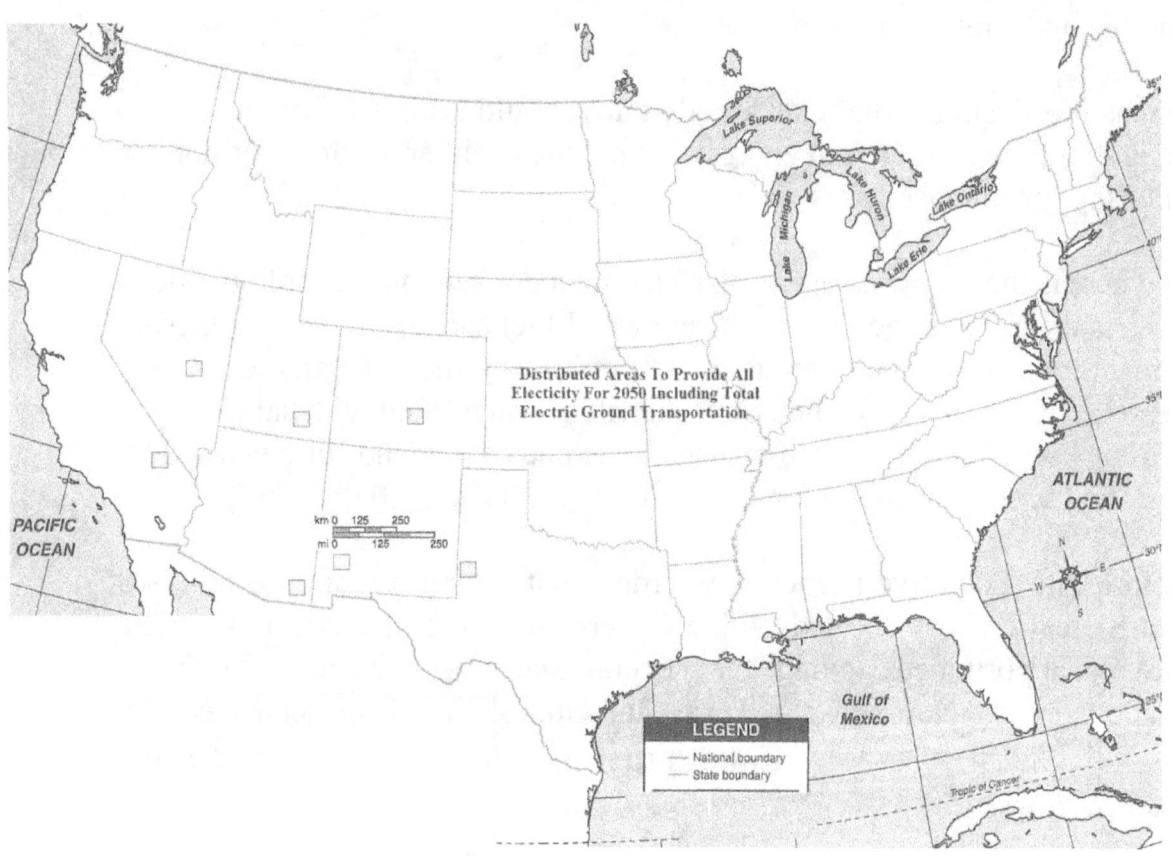

Contact Randy Ross At:

RandyRossSolar@comcast.net

703-606-2909

www.ingramcontent.com/pod-product-compliance
Lightning Source LLC
Chambersburg PA
CBHW080820170526
45158CB00009B/2481